# The Ocean World of Jacques Cousteau

## Pharaohs of the Sea

# The Ocean World of Jacques Cousteau

## Volume 9

## Pharaohs of the Sea

*Green jasper and gold, violet amethyst and silver, yellow topaz and bright red rubies proclaimed the wealth of the pharaohs of Egypt. Little did they know that in the temples of the pharaohs of the sea were jewels no less resplendent.*

The Danbury Press
A Division of Grolier Enterprises Inc.

Publisher: Robert B. Clarke

Production Supervision: William Frampton

Published by Harry N. Abrams, Inc.

Published exclusively in Canada by
Prentice-Hall of Canada, Ltd.

Revised edition—1975

Project Director: Steven Schepp

Managing Editor: Ruth Dugan
Assistant Managing Editor: Christine Names
Senior Editors: David Schulz
               Doris B. Gold
               Ralph Slayton
Assistant Editor: Jill Fairchild

Art Director and Designer: Gail Ash

Assistant to the Art Director: Martina Franz
Illustrations Editor: Howard Koslow

Production Manager: Bernard Kass

Science Consultant: Richard C. Murphy

Creative Consultant: Milton Charles

Printed in the United States of America

234567899876

LIBRARY OF CONGRESS CATALOGING
   IN PUBLICATION DATA

Cousteau, Jacques Yves.
   Pharaohs of the sea.

   (His The ocean world of Jacques Cousteau;
v. 9)
   1.  Coral reef biology.    I.   Title.
QH95.8.C68      574.5′2636      74-23073
ISBN 0-8109-0583-3

# Contents

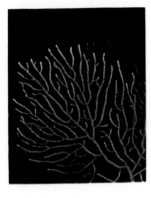

primitive plant was its main building contractor. Subsequently, throughout geological history, a number of different plants and animals took turns at being the EARLIER REEF BUILDERS (Chapter III). Four times the reef community collapsed, and each time it arose again in a somewhat different form.

To understand the reef as it is today, it may be best to begin with some understanding of the CORALS AND THEIR RELATIVES (Chapter IV). The various species take many forms in the struggle for survival. Some look like trees, some like fans, and some like pillars. Some even resemble the human brain.

To understand the coral animal, one must understand the relation it has with plants called zooxanthellae, which live with it in a mutually beneficial relationship. THE CORAL FARMER (Chapter V) makes good use of these plants, and in return it gives them protection. The zooxanthellae aid the coral's nutrition, recycle its waste products, and help it build its skeleton.

The coral reef community might be compared to a city, with its developers, its tenants, its commuters, its squatters, its food supply problems, and its waste disposal systems. It displays the most diverse and unusual assemblage of organisms in any living ecosystem. From the bottom to the top of A CORAL CITY (Chapter VI) is a succession of reef borers, cavern dwellers, and detritus feeders that include representatives of nearly every animal phylum.

There are predators on coral. In fact, a surprising number of animals feed on it. Some fish graze on its algae. Some feed on the corals themselves. Recently a certain starfish earned a sinister reputation among CORAL'S PREDATORS (Chapter VII) as a particularly voracious destroyer.

Once man lived in relative harmony with the reef. Then gradually he began to ransack its treasures. Pearl beds were exhausted. Fertilizers were shoveled away and shipped to farms in distant lands. Soils were depleted and native vegetation destroyed by vast plantations of coconut palms grown for valuable copra. Land development proceeded with too little concern for its consequences. War inflicted its wounds and so did the testing of atomic weapons. But the devastation must and can be stopped. With greater understanding and compassion, the relation between MAN AND THE CORAL REEF (Chapter VIII) could be of benefit to both.

By all means, whatever the cost, we must try to KEEP THE STONES ALIVE.

# Introduction: The Pace of Change

The very first cruise of my research ship *Calypso* was a diving expedition to the Red Sea. With some pioneer diver-biologists and geologists we explored the uncharted mazes of coral keys, islets, and snags of the Far San and Suakin reefs. It was there that I was astounded and entranced by the splendor and the folly of the coral world. But discovery and knowledge in the long run happened to trim my wildest dreams; during my deepest descent along the outer reef of Shab Suleim, sinking along a vertical wall dotted with caves and covered with brightly colored growths, I reached a sandy beach (a former sea level dating from ten or twelve thousand years ago) sloping gently to another drop off. There I was, 200 feet deep, surrounded by sharks, peering downward into an inaccessible and vertiginous dark-blue realm. Enraptured by the beauty above, challenged by the mystery below, I swore to myself that I would develop a way to quench my curiosity and descend to the roots of the reef.

Eleven years later I kept my promise and investigated many Red Sea and Indian Ocean coral reefs down to 1000 feet with a versatile exploration submarine, the diving saucer SP 350. Below 1000 feet I completed the survey with the help of automatic cameras. The picture that emerged from these investigations was very logical, but distressingly plain. The coral reefs, magnificent at the top, were huge barren mounds the size of mountain ridges covered with constantly flowing (often even cascading) sand. On flanks that looked like sloping deserts we saw big boulders that had rolled down from above. They sheltered shy goggle-eyed fish, were covered with colorful growths, and were the rare oases of a monotonous province. I had to admit the hard facts: the romantic coral reefs were gigantic tombstones crowned by a very thin mop of exuberant, obstinate, and complex living communities.

Through the diving saucer's portholes, the size of the monuments and the exiguity of the active layer of the reef suggested to me how important time had been as a construction material. During a lifetime, one witnesses a volcanic eruption, one hears of a few earthquakes, floods, tidal waves, even about the birth of an islet (which most often disappears the next week), but the map of the world changes very little. And yet, the face of the earth is constantly remodeled, but at a pace that is not ours.

One single atoll represents a volume of construction several thousand times that of the largest pyramid built by the pharaohs. And the coral reefs of the world total many million times more building stones and cement than all man's constructions throughout the ages. The industrious little polyps and the indefatigable calcareous algae have used staggering quantities of two ingredients: calcium carbonate extracted from the sea and time by the millions of years. Buried thousands of feet below the slim bustling coral cities are the fossils of the early ancestors of all reefs dating from about 2 billion years ago—almost half the life of the planet! The first reefs began to pile up their fossils long before any fish existed, and their history is a very stormy one indeed. At least four times the reefs of the world died in all oceans and remained pure funeral monuments for millions of years before conditions were favorable enough to permit their return. As did the early empires of Sumer, Carthage, or Rome, the powerful coral kingdom went through dramatic downfalls for reasons that we don't quite understand today. Each time it was reborn it knew a greater diversity and vitality. No fossil reef has ever been as rich and as beautiful as those we can study today.

To better realize the mass of time that was needed to build a deep reef, we have to use a classic analogy for distances. If you stood close to the Washington Monument, for example, and if you walked back in time 50 years for each yardlong step, two steps would be a century. After 40 steps you would have reached the birth of Jesus Christ. After 200 steps you would find yourself in prehistoric times, maybe looking for a cave. After 15 miles you would witness the very first anthropoids, our ancestors. But you would have to walk all around the earth to reach the earlier reef-building creatures! Such vertiginous explorations of time allow us to size up the difference of rhythm between man's recent hectic developments and the quiet pace of natural changes. The growing concern about the careless destruction that our unchecked technological development is spreading in the ocean is not yet fully understood. Many well-intentioned persons ask such questions as: "Insects adapt to DDT, germs to antibiotics, why would not man adapt to pesticides or to heavy metals like mercury?" Or: "Why should we feel concerned with the possible extinction of some animals as a consequence of environmental deterioration? Dinosaurs have become extinct and they have been replaced!" The answer is suggested by the patient coral community: "Because we do not allow enough time for such changes to take place. Insects adapt because of the rapid turnover of their generations. Evolution produces a very few new species every million years." If we assume that nature can cope with our feverish developments, mankind will probably meet the fate of dinosaurs.

Flying high over the Red Sea or over the Great Barrier Reef on a clear day, one is puzzled by the abundance and the variety of shapes of the coral reefs. The impression is one of eternal complexity. Unfortunately the empire of the polyps is vulnerable. If we were to bring about its fifth collapse, it may regenerate another time, probably in 10 or 20 million years. Maybe long after man himself has disappeared from the planet.

Destruction is quick and easy. Construction is slow and difficult.

Jacques-Yves Cousteau

# Chapter I. An Atoll Is Born

Around the world in a band of ocean between the latitudes of northern Florida and southern Brazil lie peculiar islands set in almost perfect rings around peaceful bodies of water. They are the coral atolls. They are especially abundant in the South Pacific. Underwater, atolls are huge annular towers that shelter some of the most vibrant and complex collectives of the ocean world. Myriads of tiny marine animals have helped to create each atoll, and now they are crowded and bustling communities of creatures that belong to nearly every major group in the animal kingdom.

> **"The corals, relatives of the sea anemone and jellyfish, are the chief architects of the reef but not its only builders."**

An atoll begins to take shape when an underwater volcano erupts and rises above the surface of warm tropical seas. If the peak of the volcano remains above or close to the surface, the larvae of simple marine animals called coral, swimming freely among the plankton of the sea, soon attach themselves in the shallow, well-lighted water along its flanks. Anchored to its new home, each microscopic, pear-shaped larva grows and secretes a limestone cup around itself. The soft part of the mature animal, the polyp, will spend the rest of its life in this external skeleton. The vase-shaped body of the polyp probably will not grow to more than a third of an inch in size. To reproduce, the polyp sends out little branchlike buds, and these grow into new polyps that remain attached to the parent. Each new polyp immediately begins to secrete its own stony cup. In time, each polyp buds in turn, and in this way a closely knit community of coral polyps gradually takes shape. They will also produce larva which settle elsewhere and initiate the formation of new colonies.

The corals, relatives of the sea anemone and the jellyfish, are the chief architects of the reef but not its only builders. Simple marine plants called coralline algae contribute by cementing the various corals together with compounds of calcium. One-celled foraminiferans donate their hard skeletons, as do tube worms, molluscs, and other animals.

The colony grows upward and outward from its point of origin upon the skeletons of corals that have died. Expanding an inch or two each year, these colonies build up the reef over the centuries. Because such a reef grows at the very edge of its volcanic neighbor, it is called a "fringing" reef. Sometimes it fails to surround the island; instead it forms a crescent-shaped reef on the windward side.

If the sea floor on which the volcano rests begins to subside, the peak that is the island also sinks. As it does, the distance between the island and the reef begins to widen. At some point the reef becomes separated by a vast body of water from the volcanic peak it once hugged. Such a reef is then called a barrier reef.

Finally, the last vestiges of the sinking peak disappear beneath the surface. If the descent of the volcano has been gradual, and if the reef has been able to grow upward fast enough, there will remain a circular coral belt all around a tranquil lagoon. Sand beaches and coconut trees turn the reef into an island. An atoll has been born.

*Volcanic birth of an island is often the beginning of a coral atoll. This island, Surtsey, near Iceland, is too far north and its waters are too cold for the growth of coral.*

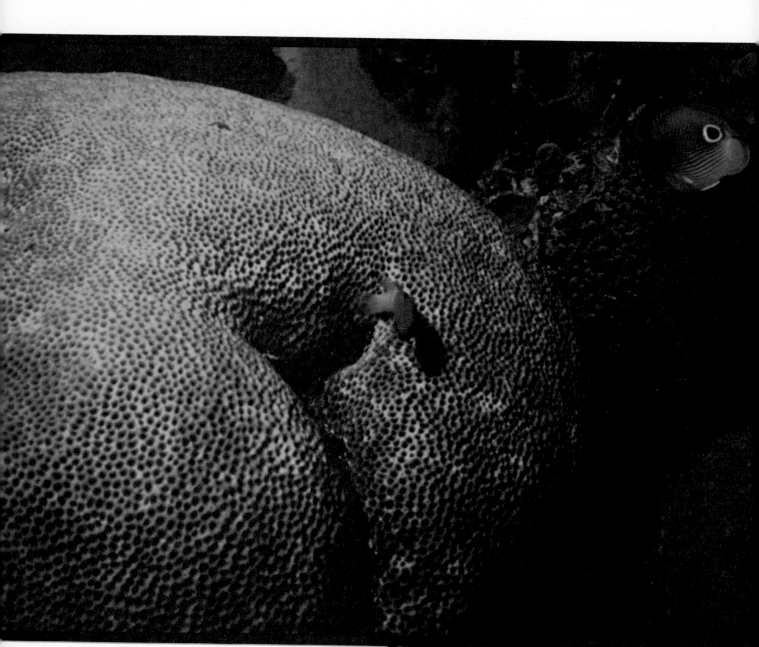

# The Builders

There are approximately 2500 species of coral and they exhibit a great diversity of colors and forms. Some are so fragile that the force of gravity causes them to crumble if they are taken out of the water.

*Reef builders. The hard corals Monastrea cavernosa, Madracis decactis (opposite page, top and center), and Siderastrea siderea (above); the sea fan Gorgonia ventalina (bottom), and a gorgonian of the genus Briareum (opposite page, right).*

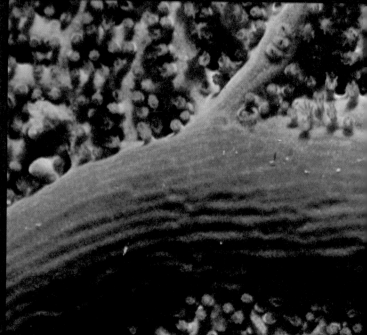

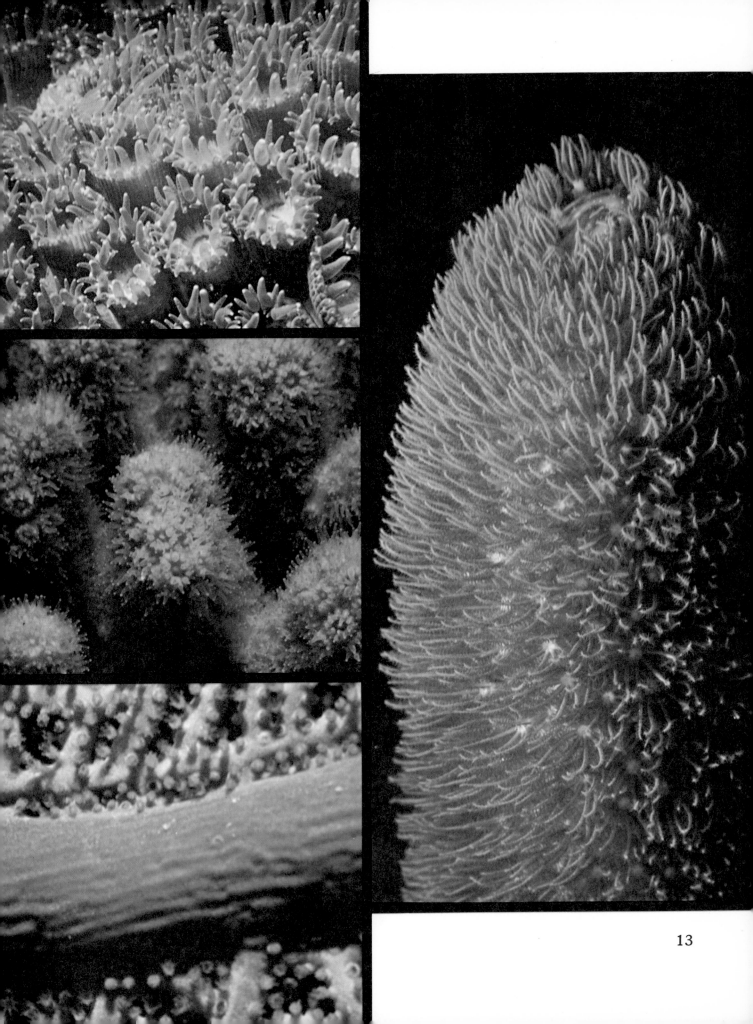

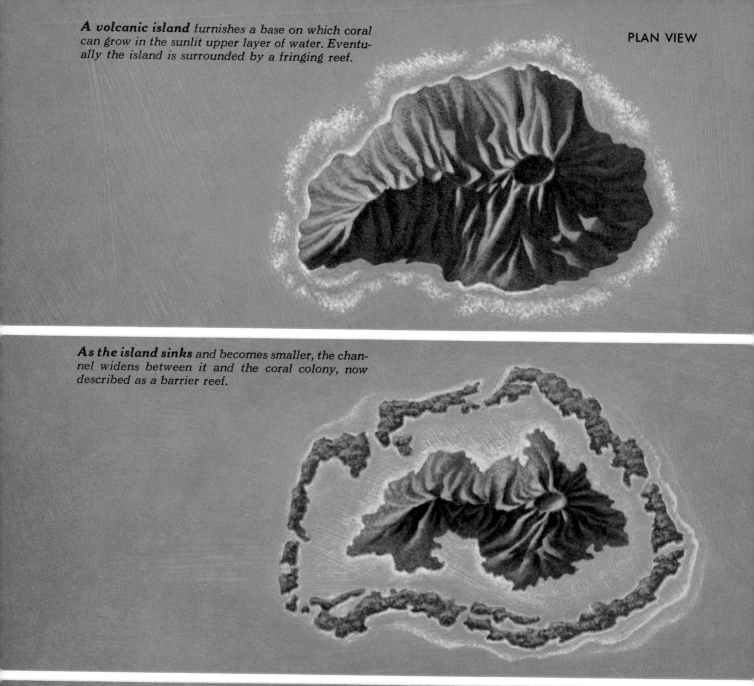

**A volcanic island** furnishes a base on which coral can grow in the sunlit upper layer of water. Eventually the island is surrounded by a fringing reef.

**As the island sinks** and becomes smaller, the channel widens between it and the coral colony, now described as a barrier reef.

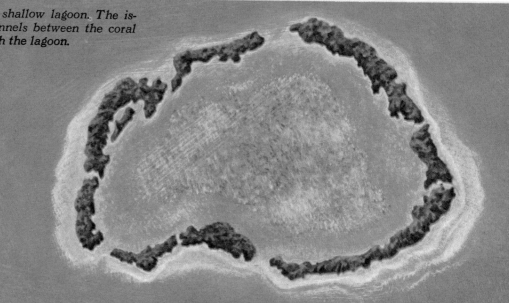

**A coral atoll** encircles a shallow lagoon. The island has disappeared. Channels between the coral islets allow seawater to flush the lagoon.

## Darwin's Vision

More than a century ago Charles Darwin, who had seen coral atolls but never explored them, correctly surmised how they were born and grew.

Other theories were presented that challenged the views of the Father of the Theory of Evolution. According to these theories, atolls were formed on shallows by an accumulation of plankton upon which colonies of coral arose. Lagoons were said to be formed because the center was cut off from the nourishment of the sea and so perished, while the coral at the rim thrived through its easy access to food from the open sea.

In 1945 a decision was made to test nuclear weapons in the Marshall Islands. Several of the group's five islands and 29 atolls were studied both before and after the weapons tests, and these investigations included deep drilling. At depths of 4158 and 4810 feet, but about two miles above the mean depth of the ocean, weathered coral and land snail fossils were found on volcanic rock. An island had once surfaced there but had settled down for almost a mile since its birth as a coral atoll. Material taken from the bottom of a 2556-foot drilling at Bikini was determined to have originated in a lagoon at a depth of less than 180 feet. On the basis of this information, Darwin's theories could no longer be questioned.

Since Darwin, scores of scientists the world over have studied how the tiny corals have been able to build structures as vast as reefs and atolls and how they were able to resist the beating of the surf and storms.

Today we know not only what coral atolls are but where they are. The orbiting celestial "eye" of NASA's Earth Resources Technology Satellite has helped supply us with information and put them all on the map.

# Supporting Cast

The coral reef is the habitat of a great variety of fishes that are remarkable for their form and color—they display almost every hue in the spectrum. Their swimming habits are adapted to the nature of the environment —corridors, mazes, and crags. •

*Reef inhabitants.* Lower photo, opposite page shows feeding sea whips, sponges, sea fans, damselfish, and striped grunts; upper photo is of a blue chromis; left is a rock beauty angelfish, and below is a fairy basslet.

# A Coral Necklace

The fringing reef surrounds the volcanic island like a necklace of coral. Only a narrow strip of water, often shallow enough to wade in, separates the reef from the island.

Recent measurements reveal that it takes from 250 to 500 years for a reef to grow seven feet. Yet it is impossible to generalize. Only a handful of the world's reefs have been measured for their rate of growth. And wave and water conditions affect the development of coral to such an extent that some reefs are receding rather than growing. At best it takes thousands of years for a fringing reef to grow around a new volcanic island. Corals need warm water to become reef builders. Even in the tropical zone, reefs will not develop in cold ocean currents. Temperature affects the chemical and physical exchanges that lead to the combination of dissolved calcium with the carbon from carbon dioxide to form calcium carbonate. Also in warm waters calcium carbonate reaches saturation in small volumes of water trapped near the coral structures and crystallizes to weld together all the reef's components.

For a coral reef to grow, not only is warm and clear water necessary, but also abundant sunlight. Microscopic, one-celled algae live within the polyp's tissues. These algae, called zooxanthellae, need the light of the sun in order to grow, just as do plants that grow out of the water. The precise relation between the polyp and the plant is not yet completely understood, but it is known that the plant helps the coral to calcify its skeleton and helps to remove the waste products of the animal.

As the fringing reef is growing, it is also being broken down. Wind, tides, and waves erode some of the reef, and the debris of plants and animals accumulates. The reef often emerges a few feet above the surface of the sea. The eroded limestone and the debris form soil in which larger plants or trees take root.

As vegetation and birds settle above the surface of the sea, the building of the reef goes on vigorously beneath the surface, where a framework of tough pinnacles of coral is formed that acts as a buffer against the sea. It is important that this limestone framework is as strong as it can be, because the reef has a tendency to grow where the surge of the sea is strongest. It is there that the open ocean is best able to bring to the coral the plankton it needs for nourishment.

*Fringing reef. Coral began to grow at the edge of this island when it emerged from the sea. A section of that reef is shown on the opposite page.*

*A circle of islets surrounds a receding volcanic cone. When the volcano eventually sinks beneath the sea, a lagoon will remain, and a true atoll will have been formed in the middle of the ocean.*

## The Living Atoll

There are hundreds of atolls dotting the South Pacific, and they are hundreds or thousands of miles from the nearest land. Atolls are relatively few in the North Pacific, extremely rare in the Atlantic, and they form clusters in the Maldive and Laccadive Islands in the Indian Ocean.

Atolls rarely close in a complete circle. They are usually a collection of islets separated by channels of water connecting a lagoon with the open sea. Darwin first called atolls lagoon islands. Lagoons would choke and die if there were no tides. The flow fills the lagoon with clean, rich ocean water, and the turbid broth of dead rotting waste is flushed back to sea through the flow channels by the ebb currents.

The bottom of a lagoon is covered with sediment formed largely from plant and animal debris and sand from the erosion of the coral rock. Here and there on the lagoon floor rise knolls of coral that sometimes reach the surface or rise above it. When they do, they are called patch reefs.

In the single lagoon of Eniwetok Atoll in the Pacific, there are over 2000 patch reefs. They dot the green or blue or olive brown of the lagoon waters with their own resplendent hues, providing a most spectacular sight. They provide good habitats for fish as well, but they can be treacherous to the navigator of the lagoon. Sometimes they are mere pinnacles of coral rock, and sometimes they are over a mile in diameter. When they grow to such a large size, they are called table reefs.

The islets displayed by the atoll are generally no more than 15 or 20 feet high. Many of them are only one or two feet above the water, and they are in constant danger of being completely inundated by the swell raised by a typhoon or by tidal waves.

There are beaches on both sides of the islets, but the sand is finer on the lagoon side. On the seaward side the beaches are usually filled with shells and fragments of coral.

Beneath the sea the abundance and vitality of coral and algae is maximum in the first 50 feet, lessening gradually to become scarce below 100 feet.

Wind, flotsam, and seabirds bring seeds to the atoll. Only hardy plants are able to sustain life on the seaward side of an atoll, splashed by salt spray and buffeted by winds. But in the relatively protected island interior, coconut and breadfruit trees are common, and morning glories grow on the lagoon beaches. Female turtles come to lay eggs, and blue-tailed skinks settle here. Crabs that eat coconuts and bats that drink fruit juice are often found on atolls. Some kinds of birds have come to stay—others make it a port of call along their migratory routes.

*A NASA satellite took this photograph of a **group of atolls**. With the aid of the orbiting camera, it has been determined that there are about 400 atolls scattered throughout the tropical seas.*

## The Windward Reef

Seen from the air, an atoll offers a dazzling sight, and also becomes easier to analyze and to understand. One of the most obvious features is that the reef is not of equal width at all points. Usually the narrow part is to leeward and the wider is to windward of the atoll, where it bears the brunt of waves and weather. It is the windward face that receives the greater amount of nutrients from the open sea to sustain vigorous growth.

Windward reefs are marked by a series of hollow grooves that cut through the upper part of the reef. These are separated by spurs of durable reef material projecting outward for as much as several hundred feet and down to depths of 100 feet. They are sometimes as high as 30 feet, and they vary in width from about 25 to 200 feet. This type

*Surge channels, barren grooves in the coral reef, serve to break the force of the pounding waves against the living structure.*

of structure is effective in dissipating wave energy without breaking up the reef itself. In some cases waves rebounding from the back of these grooves meet the oncoming wave and reduce its energy before it smashes onto the reef. The grooves also allow the water forced onto the reef to return quickly to the sea. Recent research has shown that it is the growth of the reef rather than its erosion which results in the formation of this groove and spur system. It may very well be, though, that it is the erosive action of the water as it surges back and forth over the reef that prevents growth in some places and allows it to develop relatively rapidly in others.

# Destructive Forces That Help to Build the Atoll

The moment a coral reef is born, forces set to work to destroy it. The sea begins to dissolve some of the limestone that the polyps secrete. When the reef finally rises above the surface, rain dissolves some of the stone that the seawater can no longer reach. Herbivorous fish come to grind up the coral and get at the algae. Rock-boring worms and boring barnacles make holes everywhere. This makes it easier for waves to break off big chunks of the rock and to grind it up into fine particles. Shore animals come in hordes to scrape away the rock and feed on the animal and vegetable matter that is exposed.

Waves sweeping across the rim of the atoll, transporting sediments to the lagoon, leave ripples on the surface of the reef. Even more destruction is caused when a tropical storm —a typhoon or a hurricane—unleashes its force. Many of the South Pacific reefs have had huge boulders torn away by these storms. In British Honduras after Hurricane Donna struck in 1961, 80 percent of the coral rock about five miles north of the storm's center had disappeared.

All these destructive forces work to make the finished atoll. They create the sand and sediment in which plant life can take root and where even tiny bacteria can find a home. They are also helped in the process by the life of the reef itself, for a major portion of the sediment is made up of the remains of dead plants and animals that were once a part of the living reef.

*Spurs and grooves* make up a typical pattern of coral reefs. Water returns quickly to the sea along the grooves when the waves are spent.

## Coral Sediments

Coral islands are made up almost exclusively of animal and plant skeletons. The calcareous remains of the plant *Halimeda* is one of the most numerous. The coiled shells of the tiny animals called foraminifera and the shells of many molluscs are also common. And, of course, there are the skeletons of the coralline algae and of the coral itself. The soup that erosion makes of all this is cemented together by living organisms, particularly calcareous algae and also by a precipitation of calcium carbonate.

Massive skeletons, like those of corals and algae, are resistant to mechanical breakdown, and during storms entire unbroken coral colonies can be detached and form very large contributions to the sedimentation. Breakdown of massive skeletons is less by mechanical abrasion than by biogenic erosion, or by breakdown following biological weakening of the skeleton. The branching

**Coral rubble.** *Strewn about on the beach are the remains of what was once a living reef. It was broken down by plants, animals, and pounding sea.*

coral *Acropora cervicornis* breaks down first into sticks and then into sand. *Halimeda* sediment consists of either whole plates, plates broken into two or three fragments, or sandy, silt-sized particles.

Organisms that take an important part in this breakdown include barnacles, molluscs, worms, and boring filamentous algae. Branching organisms, such as certain corals, red algae, and bryozoans, consist of segments held together by binding organic matter, which decays following death, thus releasing sediment particles. The algae *Halimeda* and *Amphiroa* are the most abundant contributors of segments, together with echinoderms, crinoids, asteroids, and ophiuroids.

Many reef organisms have internal skeletons made up of tiny spicules which disintegrate directly to fine particles upon death. These include sponges, tunicates, holothurians, alcyonaceans, and gorgonians, and some algae, especially Codiaceae.

**Jagged pinnacles.** *When the sand rose, this reef was exposed and died. Fresh rainwater dissolved the calcium carbonate in its structure.*

## Life on a Coral Atoll

Because the soil of a coral atoll is basically limestone, few species of plants can live there. Phosphates and nitrates are important to plant growth, and they are in low supply on an atoll. It takes some hardy pioneers to put down roots and flourish. The roots that they do put down must be long enough to reach the water, which does not stay near the surface of the soil but quickly sinks down to a level only a little above that of the sea. Even then, the islet must be wide enough to prevent that water from being contaminated by the salinity of seawater. Atolls in the South Pacific may have only two or three species of plants, and even if they have more, the plants are generally stunted and pale and meager in their production of fruit.

Plants are carried to an atoll as whimsically as they are carried to any other island. Seeds

Often plant life comes first to a coral island in the form of the mangrove tree. Looking like a giant spider, the red mangrove stands in shallow water protecting the leeward side of a reef, its secondary trunks circling nearby. Further inland grow the black and white varieties. This tough, adaptable tree is a fast grower, climbing about two feet in its first year, and at an early age the mangrove sends out green podlike roots that drop from the tree and hang until they become anchored.

The mangrove is able to withstand the tropical storms of its locale—Fiji, Tonga, atolls of the Indian Ocean, the reefs off Central America, Hawaii, the Galápagos Islands, Florida, and the west coast of Africa, where it is believed to have originated.

The mangrove trees help to build the island. Their roots reduce the water's flow across the reef and help accumulate sand. Droppings of transient birds enrich the soil making it capable of supporting other plants.

The variety of animals on a coral atoll is limited, too, largely because of the great distance between the atolls and other land masses; after birds, crabs are probably the most conspicuous. Insects, scorpions, centipedes, lizards, and bats are usually found. Great marine turtles can often be seen surfacing offshore and occasionally lumbering onto the land to lay their eggs. Because there are generally fewer varieties of a particular animal present, there is less competition among species with the result that populations are relatively prolific.

Most of the creatures that live on atolls share a peculiar characteristic—a tendency to dwarfism, to be smaller than other members of their species that live where food is more plentiful. An exception is one of the most spectacular beings of the South Pacific atolls —the coconut crab, the largest of its kind.

carried by ocean currents may wash ashore. Seabirds like the booby might carry them from one place to another. Or man might. Or they might be borne by the wind.

One of the seeds that most often finds its way to a coral atoll is that of the coconut. It is not a hardy seed and cannot survive long journeys, but once it has found an islet in an atoll its offspring are sure to make their way to its neighbors.

## Populating an Atoll

Many species of birds nest on atolls. For others an atoll is merely a stepping-stone on their way north or south. Some come from half the world away. Some, like the booby and the frigate bird, have never left sight of land. There are fish for them to eat, and they are even able to get their water from fish, which they desalinate within their bodies.

The birds that stay become a vital part of the ecosystem of the atoll. Consuming both plant and animal life, the birds drop their waste, rich in phosphate and known as guano. The guano can significantly alter the chemical composition of the soil, thus increasing the kinds of plants and animals that the atoll can support. The guano dropped in lagoons and in the shallows around the reef also encourages the growth of algae. And the carcass of a dead bird is an extremely rich fertilizer, useful on land and in the sea.

At nesting time, many atolls teem with birds —sooty terns and fairy terns, albatrosses and boobies. But low as it is, the atoll can be a dangerous place. Waves churned by tropi-

cal storms that lash the atoll can devastate an entire nesting colony.

The coconut crab got its name because it climbs coconut palms to eat their fruit. That is the reason it is also sometimes called the robber crab, although it never removes the coconut from the tree.

It is a giant among crabs, and it is found throughout the South Pacific and the Indian Ocean. One animal can provide a meal for several persons, and so it has been hunted to extinction on some islands.

*Coconut crab* (above) lives on lush tropical islands in the Pacific. It grows to about a foot in length; its claws are large and strong.

*Massive tern colony* (left) nests on an almost barren island in the Great Barrier Reef. The birds get their food from the sea.

It has large front claws (chelipeds) that can work through the hard shells of coconuts. Usually it does not have to climb a tree to find its meal. There are enough coconuts on the ground. But when it does climb, it does so headfirst and then comes down backward. By day it sleeps on the shady side of the tree. It does its climbing and dining at night.

The baby coconut crab develops from eggs in the lagoons or in the open sea off an atoll. Its first stages are like those of any other marine crustacean. Later it lives on land, spending about three weeks inside a borrowed mollusc shell, and then changing itself into a small land crab and burying itself in moist sand. When it emerges, it is colored light violet, deep purple, or brown.

After atomic testing in the Marshall Islands, physiologists feared that the coconut crab had become contaminated with strontium 90. But an investigation found the coconuts and crabs safe for consumption.

29

# A Profile of a Typical Coral Island

The diagrams we can draw and the descriptions we can write suggest a simplicity in the topography of a coral island that is rarely found in nature. With the development of aerial photography and with recent explorations of atolls, especially in the Pacific, the great diversity of features the islands display has become apparent. We even lack names to describe some of the more complex forms that have been discovered. Nevertheless, we are able to point out the most distinctive features that are common to every fully developed coral atoll.

Hidden beneath the surface of the sea and beneath layers of dead coral and sediment lies the old volcano that once was an island but has long since returned, like Triton, to its watery grave. Above this volcanic pedestal, lying in the center of the ring of islets, is the legendary lagoon, seldom more than 200 or 300 feet deep. In the deeper central waters of the lagoon, pillarlike coral knolls

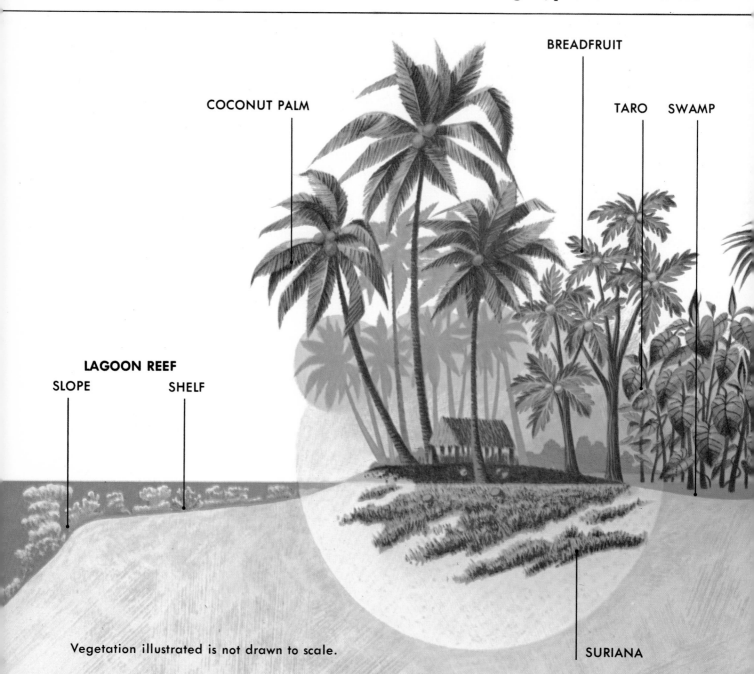

BREADFRUIT

COCONUT PALM

TARO    SWAMP

LAGOON REEF

SLOPE    SHELF

Vegetation illustrated is not drawn to scale.

SURIANA

rise from the floor. Along its rim stand tide pools among limestone boulders, sand flats often overgrown with eelgrass, and patches of branching hard corals. Deep channels between the atoll's islets connect the lagoon with the open sea.

The windward side of the atoll is more fully developed than the leeward side. The reef flat is broader. Its seaward front usually has developed an intricate underwater system of grooves and spurs for its own protection, and there are beaches on either side of the reef. Leading from the lagoon to the windward islets, the beach is generally colored tan or orange. On the seaward side of the windward islets the beach is rougher, with far more large pieces of rubble.

Sometimes the islets are rich in vegetation. Sometimes they are able to support only shallow scrubgrasses in their barren limestone soil.

At its outer edge, the steep seaward reef slopes off into the depths of the ocean, the boundary of one of the sea's most complex and curious constructions.

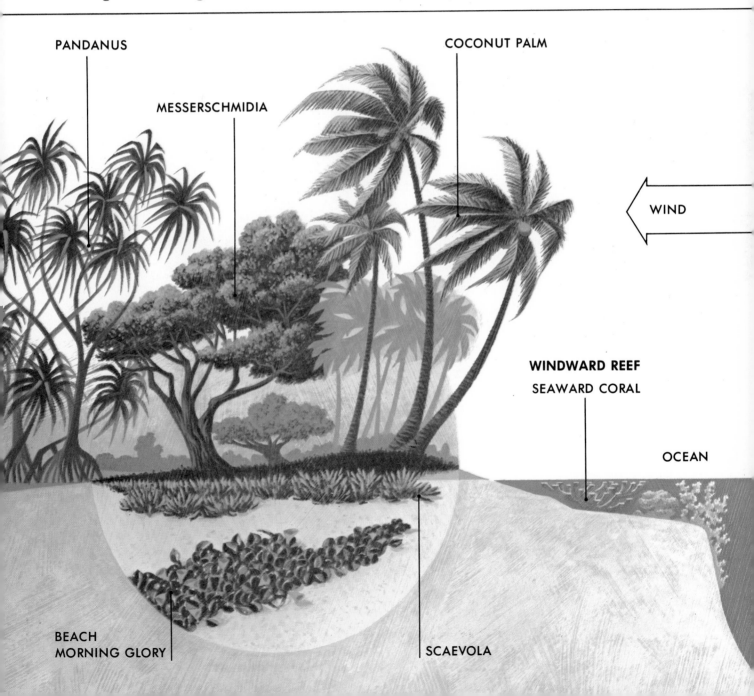

PANDANUS

MESSERSCHMIDIA

COCONUT PALM

WIND

WINDWARD REEF

SEAWARD CORAL

OCEAN

BEACH
MORNING GLORY

SCAEVOLA

# An Island That Rose Again

Clipperton Island, an atoll 600 miles southwest of the Mexican coast off Acapulco, is a typical atoll, remarkable because its lagoon is closed and filled with slightly brackish fresh water. But that lagoon was once connected to the open sea. Clipperton is an island that sank and rose again, and we are able to date the lagoon's closing accurately.

In 1837 a hydrographer of the British Navy drew the first known map of the island and showed it to be discontinuous, interrupted by two channels that gave onto the open sea. Scientific missions to Clipperton in the 1960s found it with a lagoon completely enclosed and capped with a layer of brackish water. Clipperton has risen and retained the rainwater fallen within.

Clipperton is seven miles in circumference with a 16-foot coral crown above the water. At one end of the atoll, a single volcanic rock rises to 94 feet above sea level. The surface of the atoll is generally bare, its dead coral buried only by sand or lichens. Some parts of the atoll have a brilliant green ivy cover. Others are shaded by a few coconut palms.

Clipperton is covered by a "faultless vermilion red carpet" of red land crabs known only on Clipperton, Costa Rica, Baja California, and two islands off Colombia.

Birds abound on the island. Its estimated population of 26,000 is comprised of more than 100 species, including the yellow-beaked cuckoo and the cattle-herd heron, two species that were discovered here.

Clipperton's underwater shelf is swollen and beaten by the Pacific. Landing there is almost impossible except by helicopter.

*A narrow ring* of sand surrounds the closed and stagnant lagoon of Clipperton, a volcanic island that sank and rose again.

# Chapter II. The Great Barrier Reef

A barrier reef is like a fortress of rock. Often it rises steeply on its windward side, a great distance separating it from the island it seems to guard. It may be a continuous strip of land, or it may be a series of smaller patches running closely together.

---

**"The Great Barrier Reef might be considered the eighth natural wonder of the world."**

---

The most spectacular of the barrier reefs is the Great Barrier Reef. This immense coral land strip was first navigated by the British explorer Captain Cook in 1768-69. It was explored scientifically by marine geologists in 1928-29. It is 1260 miles long and from 10 to 90 miles wide. It lies from 10 to 100 miles off the Queensland coast of Australia. It is made up of 200 coral-formed sandkeys and thousands of islands with fringing coral reefs. It is the habitat of countless species of marine creatures.

Thanks to its favorable climatic location, to its deep shelf, and to the high prevailing tides (9 to 11 feet) that generate strong nutritive currents, the reef has flourished for many thousands of years. It might be considered the eighth natural wonder of the world.

Both the three major wind systems and the temperature of water in the tropical southwest Pacific are congenial to the development and sustenance of coral reefs. From May to November the trade winds prevail in the northern reef areas, but in the Australian summer—December to April—steady high-velocity winds cause strong surface drift. During the same period, very strong monsoon winds blow for four months.

The three major water masses of the area are the equatorial, central, and subantarctic.

Their temperature varies from freezing in the subantarctic to 80° F. in equatorial waters. Where there is a high degree of salinity, as in the warm Coral Sea, a high carbonate ion concentration is found. Carbonate compounds are required by those sea organisms that secrete limestone—the building block of the reef.

The Great Barrier Reef rests on major oceanic rises and ridges on the continental slope and is a shelf reef. In his ship's log Captain Cook mentioned "a wall of Coral Rock rising almost perpendicular out of the unfathomable ocean." The reefs were treacherous to sailing ships of Cook's time, and his *Endeavour* ran aground. He was forced to throw 50 tons of cargo overboard to free the ship. Philadelphia Academy of Natural Science divers recovered some of Cook's coral encrusted cannons in 1969.

There are no atolls in the Great Barrier system, but there are many platform reefs, patch reefs, boulders, and pronglike or triangular areas of coral beaches between the Australian continent and the reef itself. The reef was first fully mapped from 120 miles in space by the crew of Apollo VII.

In the reef there are bodies of water 1.6 times the size of the entire land area of Australia. Differing wind, wave, and water conditions have created a variety of environments along the length of the reef. Because of its staggering extension, the Great Barrier Reef offers an almost complete catalog of coral reef populations.

*Mosaic of corals makes up the 1260-mile-long Great Barrier Reef of Australia, the largest barrier reef on earth. It has thousands of islands, platform and fringing reefs, but no atolls.*

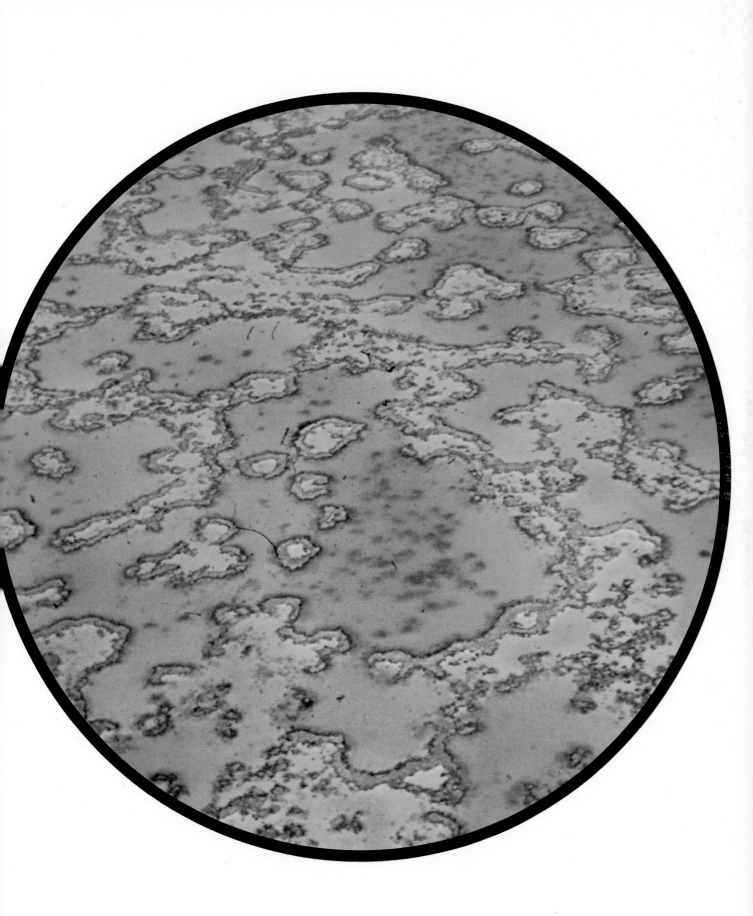

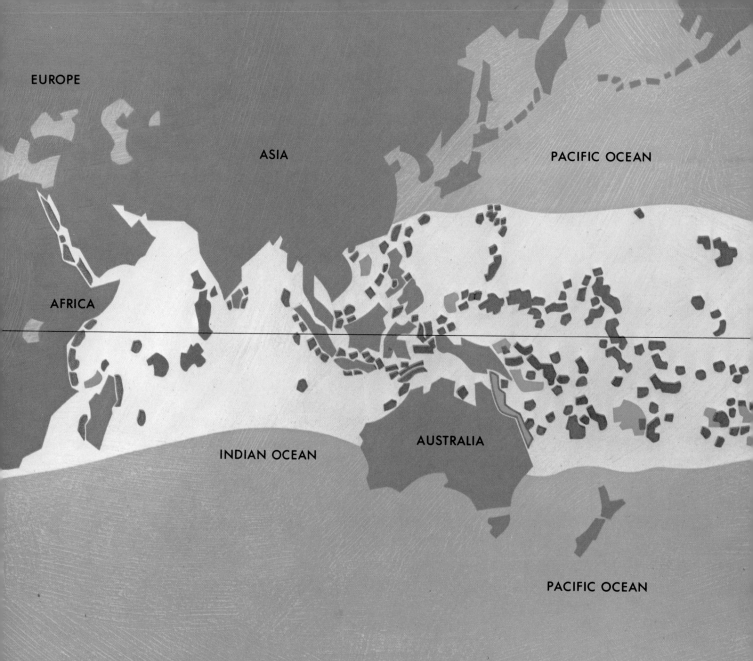

EUROPE

ASIA

PACIFIC OCEAN

AFRICA

INDIAN OCEAN

AUSTRALIA

PACIFIC OCEAN

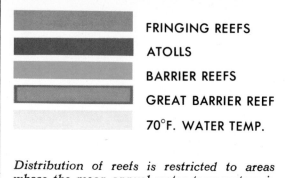

FRINGING REEFS

ATOLLS

BARRIER REEFS

GREAT BARRIER REEF

70°F. WATER TEMP.

*Distribution of reefs is restricted to areas where the mean annual water temperature is 70° F. None exists in the eastern Atlantic.*

## Coral Reefs Around the World

Coral reefs are scattered in the Pacific, the Caribbean, the Indian Ocean, and the Red Sea. There are a few minor reefs in the South Atlantic, at Bermuda, in the Persian Gulf, and along the tropical eastern coastline of the Americas.

There is a distinct relation between the occurrence of atolls in the Pacific and the "ring of fire," a zone that skirts the edge of that ocean along major fault lines in

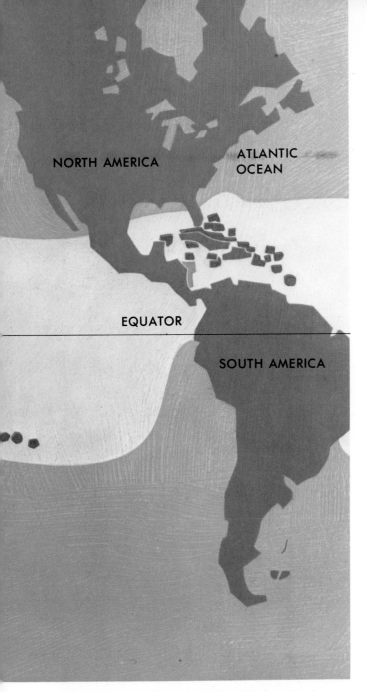

NORTH AMERICA

ATLANTIC OCEAN

EQUATOR

SOUTH AMERICA

the earth's surface and is subject to earthquakes as well as volcanic eruptions.

Nearly all the smaller islands of the Pacific that are far from the continents are of volcanic origin. Some are exposed summits of volcanoes, and of these some have fringing reefs around them. Others are atolls.

There are about 10,000 volcanoes in the Pacific Ocean, most of them hidden beneath the sea and no longer active. One of the active volcanoes is about 750 miles southeast

of Tahiti. Ten thousand feet above the sea floor, its summit still lies 1500 feet below the surface of the ocean, and it continues to build upward. Some day this volcano could erupt and rise above the surface of the sea. If it did, it could mark the birth of yet another coral reef.

Kwajalein in the Pacific, 78 miles long and almost a third as wide, is the world's largest atoll. The smallest are the lagoon patch reefs that are referred to as microatolls.

The distinction among fringing, barrier, and atoll reefs has been extended since Darwin named them to include other stages in development, and we now recognize apron reefs, which are embryonic fringing reefs, and table reefs, which are open-ocean reefs without either a central island or lagoon. Intermediate stages in the development of reefs are referred to as almost barriers, almost atolls, and almost tables.

The greatest diversity of types of coral is in the Indian and Pacific oceans. In the Melanesian-Southeast Asian area there are over 50 genera and 700 species. Most of the Indian and Pacific reefs have from 20 to 40 genera. In the Atlantic there is far less diversity. The greatest variety is in the reefs of Puerto Rico and Jamaica, where 26 genera and 35 species have been counted. The world's northernmost atolls, Kure and Midway in the Pacific, have only nine genera, compared with 52 in the Marshall Islands.

Some of the coral islands became familiar names during World War II—Guam, Saipan, Midway. So did some of the groups of islands—the Carolines, the Marianas, the Marshalls, and the Gilberts. Some of the coral reefs are familiar because they are close to tourist islands—Hawaii, Jamaica, the Bahamas, Curacao, Barbados, and Bermuda. A few, like Bikini, became famous as the sites for the testing of nuclear weapons.

*The sponge,* unable to move itself, strains its meal from the great amount of water it passes through the minute pores of its body.

## Sponges

Several limestone-secreting families of sponges are next in significance as reef-builders to the corals and to calcareous algae. Sponges are known as *Porifera,* and the name means "pore bearers." Their weird shapes and bright colors extend the beauty of the reefs deeper than coral can thrive; but they are very primitive: they have no organs, no mouth or digestive tract, no nervous system. Their body is permeated with pores, canals, and chambers through which sea-water flows. Functions are performed by the cells acting more or less independently and cooperating little with each other. Most sponges grow in irregular shapes, forming flat, rounded, branching, or encrusting structures. Sponges get their food from minute organisms drifting in the water that they take in through the thousands of openings covering their surface. Nearly all sponges possess an internal skeleton. Mature sponges are incapable of movement, and they live fastened to rocks, shells, and other objects. Sometimes, but only rarely, they roll about on the bottom of the sea.

Calcareous sponges have spicules that are largely crystalline calcium carbonate. Near the surface they are generally quite small, ranging from about an eighth of an inch to five inches in length, but deeper along the flanks of the reef some sponges may stand three feet high.

*A **soft coral** of the Great Barrier Reef* (Sarcophyton trocheliophorum). *The skeleton consists of hard spicules in a rubbery rind.*

## Soft Corals of the Reef

The soft corals, or alcyonaceans, have only scattered spicules to stiffen their bodies. Although some soft corals extend into polar waters, they are largely a warm-water group, most abundant in the Indian and Pacific oceans. Some are shaped like leathery brown, great flabby masses; others are like pink or purple branches of celery.

Two relatives of soft corals which also do not build reefs are the red organ-pipe coral *Tubipora* and the blue *Heliopora*. The organ-pipe is found on the reef flat. Blue coral is found on the reef front and slope.

The organ-pipe coral has emerald green polyps that emerge from brick-red limestone tubes. Their massive layers of tubes are not secreted as solid limestone but consist of fused spicules and lie within the living tissue of the coral. The tubes are joined at intervals by platforms in which run connecting digestive canals. As a colony grows, the lower levels of the tubes are abandoned, and they become the habitat of worms, small crabs, and countless other little animals.

The blue color of *Heliopora*'s skeleton is thought to be due to biliverdin (a pigment), but the "blue" coral does not look blue at all when its brown polyps are extended.

These animals, like the hard corals, are colonial, but instead of having six unbranched tentacles, they possess eight delicately branched tentacles on each polyp.

39

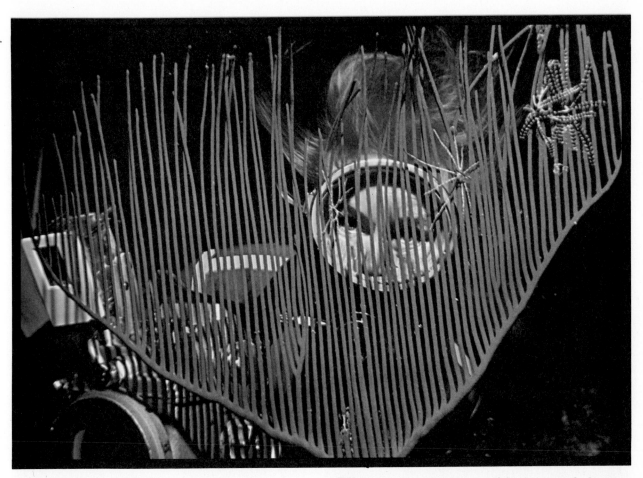

## Great Barrier Reef Corals

Branching corals are perhaps the most abundant group to be found in the Great Barrier Reef of Australia. They take many shapes: some look like large stone-lace dishes; others like garden fences, tufted clumps, or bushes. Some have short, stubby branches; others produce thin, long, stony spinneys and brakes. They look fragile but can resist severe wave action. Many branching corals reach their maximum development in the living coral zone behind the algal rim and on the reef front.

Brain corals grow in the middle and inner reef flat where they form small atolls. They are massive, rounded structures of great strength. They resemble a human brain, and they range in diameter from a few inches to more than two feet.

Plate corals assume a wide range of shapes, but as a group they are characterized by their thin, laminar skeleton which may be flat or folded to produce intricate, foliate patterns. They are mainly found on the reef front, but some genera also flourish in various parts of the reef flat.

Encrusting corals inhabit the outer reef flat and the reef front. They grow on other organisms and on debris in the reef that are literally enveloped by them.

Mushroom corals are single-polyp forms with large central cavities. They are seen mostly in the coral pools or moats of the outer reef and are free to move.

And then there are other kinds of coral that belong to none of these types. Some relatives are flexible, like the gorgonians. Others, like the fire corals, are hard.

## Gastropods

Some of the most colorful animals of the reef are the gastropods—snails and sea slugs called nudibranchs. They are molluscs, akin to the squid, the clam, and the abalone.

Snails characteristically have single trochospiral (coiled, spiraled, and elongate) shells. They move on a single large "foot," which gives the appearance of having originated from the stomach (hence "gastropod"). Snails seem to be plodding, gentle creatures and they are—except when feeding. The snail's tongue is covered with filelike teeth used to rasp away vegetable or animal matter. Some snails use this tongue to drill shells of oysters and clams for food.

The nudibranch is a snail with a shell that disappears shortly after the little animal has passed its larval stage. Plumelike projections on the back or surrounding the anal opening give these creatures their name, for they are the naked gills by which nudibranchs breathe.

*Sea slug on coral* (left) when fully grown will reach approximately three inches in length.

*Shell-less nudibranch* (right) displays its respiratory gills in a crown that circles anal opening.

*Brightly colored gastropod* (below) adds to the spectacular beauty of the Great Barrier Reef.

*A **crinoid** clings to a red gorgonian coral. Graceful swimmers, these animals are often found attached to solid objects on the bottom.*

## Crinoids

A crinoid is an animal with delicate branching arms that form brilliantly colored bouquets in clear water on coral reefs just below the tide line. *Crinoidea* means "lilyform," and the name was well chosen. They are commonly known as sea lilies or feather stars.

The crinoid's body is disc-shaped, and it has feathery arms. It can grow new arms or even a new central disc if the original one is lost. For food they depend on small animals and plants that drift by on the current. Crinoids snare this prey by reaching out with arms that are usually fringed along both sides with a row of short, tapering branches. Most

*A **feather star** lifts fringed arms to catch small particles of food that are moved to mucus-filled channels and transported to the mouth.*

crinoids have five arms, which are branched at their base, but some have as many as 200 arms.

Feather stars establish themselves on the bottom with a slender stem. Later they break away from its upper end and ever after lead a free existence. Feather stars that live where abundant sunlight can reach them are bright red, orange, golden yellow, white, black, purple, or variegated; their colors are perhaps more beautiful than those of any other marine animal.

Many limestone beds contain crinoid fossils from the Paleozoic era, when they were extremely successful. Feather stars are the only crinoids found on the Great Barrier Reef.

## At Home on the Reef

The Great Barrier Reef is the home of a wide variety of marine animals. There are molluscs and shelled amoebas (foraminiferans), stars, brittle stars, urchins, cucumbers (echinoderms), sea squirts (ascidians), crustaceans, and polychaete worms.

Among the most colorful of the reef inhabitants are the brittle stars found on the reef surface and in deeper waters beyond the reef. They have long, extremely flexible, snakelike arms that they will break off to escape from predators. They often coil their arms around branches of coral like minute pythons in a tree.

Sea cucumbers (holothurians) abound on the reef. Some are black and soft-skinned, while others are larger, brown-spotted, and rough-skinned. There are extremely long,

*A basket star uncurls its multibranched arms to feed in the dark nighttime waters.*

wormlike cucumbers with a ring of tentacles at their mouth. Some cucumbers are considered a delicacy by Asiatic gourmets.

There is a black, long-spined sea urchin that lives in groups in the coral pools of the outer reef flat. Another urchin is a borer with short, strong, sharp spines. It is common on the algal rims where it grinds out spherical hollows in which it lives. The slate-pencil urchin has thick, blunt, penlike spines.

**Sea fan's** *intricate network of branches partially hides and protects these long-nosed birdfish.*

Sea squirts an inch and a half tall placidly filter plankton, their food, from the sea. Giant clams that weight a quarter of a ton open velvet-lined vises four feet wide that cradle its cuplike siphons.

All of these animals and many more make some form of contribution to the vitality of the reef community in their lifetime, and most of them add something to the growth of the reef with their death.

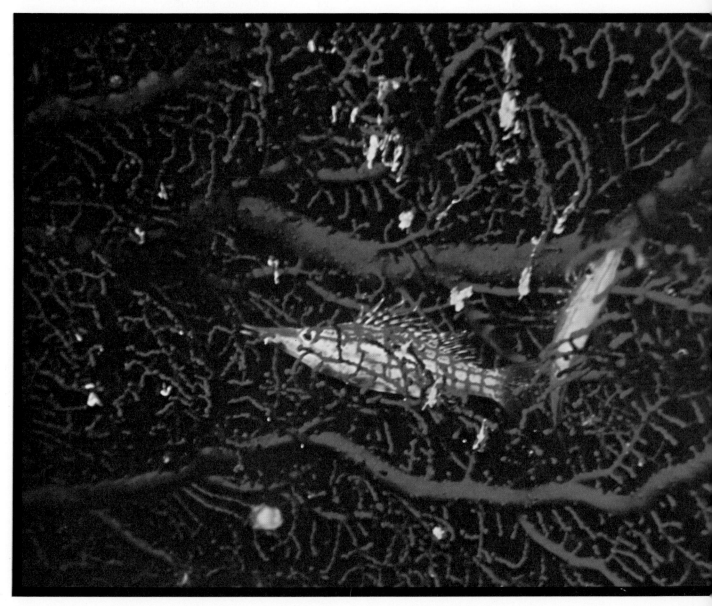

# Great Barrier Reef Fish

The Great Barrier Reef of Australia has about 1500 species of fish. There are 28 kinds of sharks and 30 kinds of rays, including the spectacular electric torpedo ray and the manta ray. The eagle ray pounces on crustaceans and clams and crushes their shells with its powerful flat tooth plates. Rays may grow up to a fin span of seven feet. Migrating open-ocean fish, like tuna, bonitos, and sprats, pause around the reefs as an attractive port of call to break their monotonous journey.

The reef giants include groupers, barracudas, coral trouts, and moray eels.

*Underwater garden* of corals and sponges (left) is home to loosely schooling blue chromis and brightly colored royal grammas.

*Big-eye* (below) peers from its retreat in the side of a coral reef. It will withdraw into its hole completely at any sign of danger.

*Matching colors* of the reef is a toadfish (above). By remaining motionless, the toadfish attracts no attention from either predator or prey.

*Sharing the reef* (right) are many kinds of fish and invertebrates. Among them are the yellow-and-black rock beauty, striped grunts, and blue chromis.

The Great Barrier Reef has at least its share of dangerous fish. The stonefish, the lionfish, the butterfly cod, the frogfish, the stingray, the sea snakes, and the catfish use their poison mainly for defense. The stingray has a serrate spine at the base of its tail that is not only poisonous but can cause a painful and dangerous wound.

Among the smaller crowd, the clownfish is a brightly colored little creature that lives safely with its venomous host, the large sea anemone. Perhaps the most intriguing fish of the reef are the seahorses and their relatives —the hobbyhorses, weedy sea dragons, ghost pipefish, and razorfish.

# Chapter III. Earlier Reef Builders

The coral reefs we know are relatively recent structures. There have been other, earlier reefs. The first, and simplest, were created 2 billion years ago in the middle and late Precambrian times by algae working without assistance of animals. Geologists call them stromatolites, from the Greek words for "flat" and "stone."

The stromatolite reefs were enriched about 600 million years ago by stony, spongelike animals called archaeocyathids, Greek for "ancient" and "cup." The archaeocyathids disappeared 540 million years ago in the first collapse of the reefs. Algae continued simple building activity without a partner for the next 60 million years.

A complex new building cooperative formed some 480 million years ago. Fossil evidence identifies its members: stromatolites, coralline red algae, colonial bryozoans, stony sponges, and the first of the corals. This third type of reef lasted approximately 130 million years, during which the special relationship between algae and coral was established. Sponges that were to become fundamental partners in ocean reef construction down the ages emerged in both encrusting and free-standing varieties.

---

## "Throughout geological history, the reef communities collapsed four times."

---

Around 350 million years ago all of the reef builders, save the persistent algal stromatolites, were killed off in the second collapse. It was 13 million years before reef development began again. This time two new groups of calcareous green algae joined the stromatolites. Bryozoans again became prominent and chambered sponges appeared.

For about 115 million years the reefs evolved until, about 225 million years ago a third and devastating collapse occurred. Half the known marine and land animals

*Rise and fall of the reefs.* In the 2 billion years since the first reef building began, four major collapses have occurred. They are believed the result of environmental changes. The third collapse coincided with the extinction of dinosaurs.

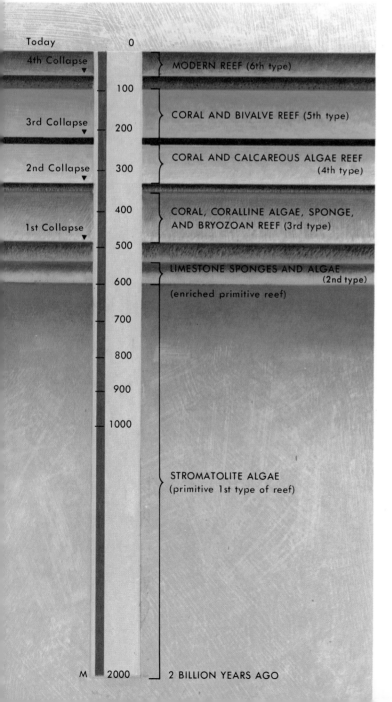

Today — 0
4th Collapse
— 100
MODERN REEF (6th type)
3rd Collapse — 200
CORAL AND BIVALVE REEF (5th type)
2nd Collapse — 300
CORAL AND CALCAREOUS ALGAE REEF (4th type)
— 400
CORAL, CORALLINE ALGAE, SPONGE, AND BRYOZOAN REEF (3rd type)
1st Collapse — 500
LIMESTONE SPONGES AND ALGAE (2nd type)
— 600
(enriched primitive reef)
— 700
— 800
— 900
— 1000
STROMATOLITE ALGAE (primitive 1st type of reef)
M — 2000 — 2 BILLION YEARS AGO

and plants became extinct. Reef builders did not reappear for 10 million years. Then corals made a comeback, this time as six families in the scleractinian group, from which the families of corals we know today have descended.

In the ensuing 130 million years reefs spread from a few scattered patches to flourishing settlements in many parts of the world. Over the period the reef's inhabitants changed. New groups of sponges, sea urchins, foraminifera, and molluscs made their appearance. Roles changed too. Stromatolites faded while a coralline red algae took over the partnership with an increasingly diverse number of coral families. A challenge to the dominant corals came from molluscs called rudists that left fossil aggregations of shells cemented together. Suddenly, the reefs suffered their fourth collapse. The rudists were wiped out, and two-thirds of the coral families disappeared. This happened 65 million years ago, at the time dinosaurs became extinct. After 10 million years the reefs began to take hold once more. They grew vigorously and have maintained themselves to the present time.

The dramatic reef crashes are believed to be connected with cataclysmic periods in the history of the earth itself. Over 2 billion years vast changes have taken place; ocean basins formed, continents came together and broke apart, climates that were moist and mild turned severe and dry. Vast though the changes were, they were gradual.

*Fossils* *from a reef of the Devonian period of more than 350 million years ago include a number of bryozoans and crinoids. Some of the other builders of this third type of reef were stromatoporoid sponges, tabulate and rugose corals, and coralline algae.*

The challenge the reefs face today is more sinister. Professional divers and diving biologists have become concerned in the past 20 years that coral vitality is decreasing rapidly throughout the world. Reefs are destroyed systematically with crowbars to supply novelty shops and museums with coral and shells. Spearfishermen deplete the reef's fish. Pollution already stretches its grim threat to the most remote oceans.

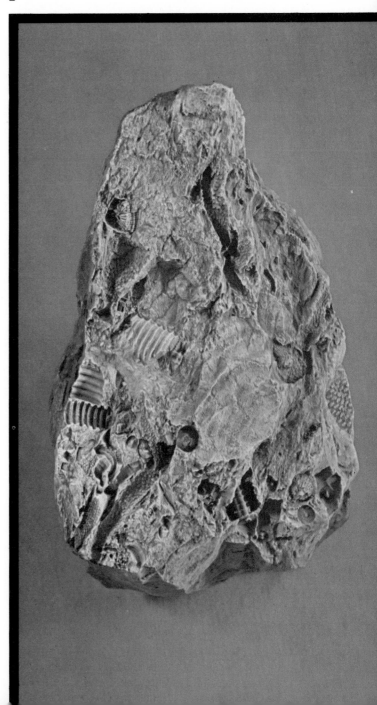

## The Role of Algae

Algae were the builders of the earliest reefs. These small and fragile organisms are rarely preserved as fossils, but they were probably similar to the blue green algae that form similar masses of limestone today. The accomplishments of the Precambrian algal reef builders were not inconsiderable: individual colonies grew upwards for tens of feet. Fossils of the structures they built are in the form of trunklike columns or hemispherical mounds.

The close association between algae and corals that apparently began 480 million years ago has continued (albeit with notable fluctuations) to our day.

During later Carboniferous and Permian times, from about 325 to 225 million years ago, two new groups of calcareous green algae, the dasycladaceans and codiaceans, appeared and attained quantitative importance in the reef assemblage.

*Encrusting purple calcareous algae act as a cement, binding reef fragments together.*

*Green algae,* **Halimeda.** *Its calcified tissue constitutes one of the many sources of sand.*

During the 130 million years or so of Jurassic and Cretaceous times, a hitherto minor group of coralline red algae, the lithothamnions, began to play an increasingly important role, such as building "sidewalks" all along the reefs just below the sea level.

In our own time algae—blue green, green, and red—are the principal food base of the reef community. The blue green algae are shallow tidal flat organisms, and the reds are mostly fore-reef inhabitants.

Some of the encrusting calcareous algae have interesting fluorescent properties.

Divers 200 feet deep along the steep reefs, looking closely at the cliff, can see patches of bright yellow, orange, or red in the blue ambient obscurity. If they carefully separate the thin algal plates and bring them to the bright light of the surface, they are just dull brown. These algae absorb the luminous energy in the violet and blue band widths and reemit some of the energy in another wavelength corresponding to the bright colors seen by the divers.

## Early Reef Builders

A spongelike creature, the archaecocyathid, was the first animal to join in the making of reefs, almost 600 million years ago.

For millions of years in mid-Ordovician times true stony sponges, called stromatoporoids, enriched the growing community of reefs. Some were shaped like encrusting plates; others were hemispherical or shrublike.

At the beginning of the Permian era appeared a new type of sponge, calcareous and chambered, the sphinctozoan.

We have mentioned earlier that the Paleozoic era ended about 225 million years ago with the extinction of half the known taxonomic families of animals, both terrestrial and marine, and a large number of terrestrial plants. In the reef community a successful association among algae, bryozoans, and the sphinctozoan sponges came to an end. Reefs were unknown anywhere in the world for the next 10 million years.

By the Jurassic era, when the stromatoporoid sponges reappeared in the Mediterranean extremity of the Tethys Sea, these sponges and the chambered sponges coexisted. Ten million years after the close of the

Cretaceous period which coincided with the fourth collapse of all reefs, a new period of growth began (near the end of the Paleocene epoch). Stromatoporoid sponges were a part of the revival and they have flourished substantially unchanged to the present day.

This brief outline shows that sponges were closely associated with the rise and fall of the reefs through the ages. The drama extends for hundreds of millions of years, and it is practically impossible for the human mind to realize how slow all these transformations were. We are now modifying our planet (and spreading destruction all over it) at a speed that nature is totally incapable of coping with. Natural evolution and man-made evolution do not use a similar timepiece. If mankind were to disappear in a global catastrophe, sponges would have a good chance of reappearing 10 million years later.

*A brittle star* (above, left) wraps its arms around a sponge. Coral to the right of the sponge is Monastrea cavernosa, *an important reef builder.*

*Many-pored sponges* (right) may grow to a foot tall. Some species become large enough to seat an adult man. Others are merely thin encrustations.

*Pumping water* through their pores, primitive sponges (below) have flourished with the reefs over hundreds of millions of years.

## The Moss Animals of the Reef

There are today about 3000 species of marine moss animals (bryozoans) ideally suited for life on a reef. They attach themselves to the sea bottom and feed on microscopic animals caught in their tentacles. Superficially their skeletons look like coral skeletons, but they are smaller and lack mesenteries. Geologists look for them as evidences of past construction activity.

During the Permian period, about 250 million years ago, they were among the reef builders. Even earlier, in middle Ordovician times, about 480 million years ago, bryozoans became important contributors, together with red algae and sponges, in the development of reef rock.

So minute are they that thousands growing in colonies were needed to form a structure several inches across. Many secrete a skeleton of calcium carbonate, making them perfect reef inhabitants. Their skeletons display various shapes, some branching like small trees, others encrusting, still others having the appearance of convoluted lace.

*Lacy bryozoans* of western Samoa dwell in rocky crevices. They secrete calcium carbonate skeletons that contribute to the reef's structure.

# The Role of Molluscs

The skeletons and shells of molluscs—clams and oysters; octopi and squids; snails, scallops, and chitons—are stones, bricks, and cement for the ocean's huge construction workshop.

The reefs that developed in the Tethys Sea, the wide tropical seaway of late Paleozoic times that almost circled the globe, included various molluscs.

In the Cretaceous period certain previously obscure bivalves known as rudists went through a period of phenomenal proliferation, and during the next 60 million years they reached a position approaching dominance among reef animals. Along the sheltered landward margins of many fringing and barrier structures, rudists largely supplanted the corals. Their cylindrical and conical shells were cemented into tightly packed aggregates, and many of these grew upward in columns that imitated some coral growth pattern. At the end of the Cretaceous period some 65 million years ago, the rudists quite abruptly died out everywhere.

*Orange-mantled scallop displays its rows of bright eyes. Shells of molluscs cemented together formed reefs more than 65 million years ago.*

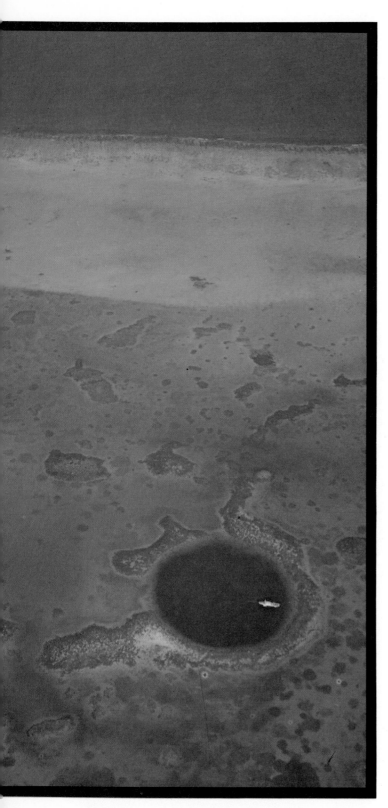

*Blue hole of Lighthouse Reef.* Calypso *divers explored this deep pit in the barrier reef off British Honduras and found evidence at 140 feet that it had once been above water.*

## The Sunken Caves

Some of the strangest geological phenomena on earth are mysterious holes in shallow table reefs. Because of their depth, the water they contain is a deeper blue than the sea around them. So they are called blue holes. Divers of the *Calypso* explored the largest known blue hole, that of Lighthouse Reef off the coast of British Honduras. At a depth of 140 feet they explored a gigantic cavern containing stalactites. The blue hole of Lighthouse Reef had once been above sea level!

Stalactites and stalagmites are formed in caves as rainwater drips from the ceiling after filtering through the ground. They cannot be built underwater. The stalactites were formed during centuries in air, and the cavern was drowned when glaciers of the ice age melted and the oceans rose about 12,000 years ago. The enormous barrier reef of the Caribbean, second in size only to Australia's Great Barrier Reef, dates from that time.

More than 100 blue holes dot the waters around Andros Island in the Bahamas. Here too, George Benjamin discovered stalactites in a cave 169 feet under the water. Suspecting that the holes and tunnels might be interconnected, the divers poured a harmless green fluorescent dye into one of the holes. Soon the green dye was seen escaping from many holes and vents in the reef, proving that Andros and its reef are riddled like a giant sponge.

The divers noticed something more about the sunken tunnels of Andros. Seawater going into them was murky with plankton and organic material. When the tidal flow reversed, water coming out was crystal clear. In one dark tunnel a diver discovered a rare gorgonian growing without sunlight, obtaining its energy by feeding on plankton brought to it by the currents.

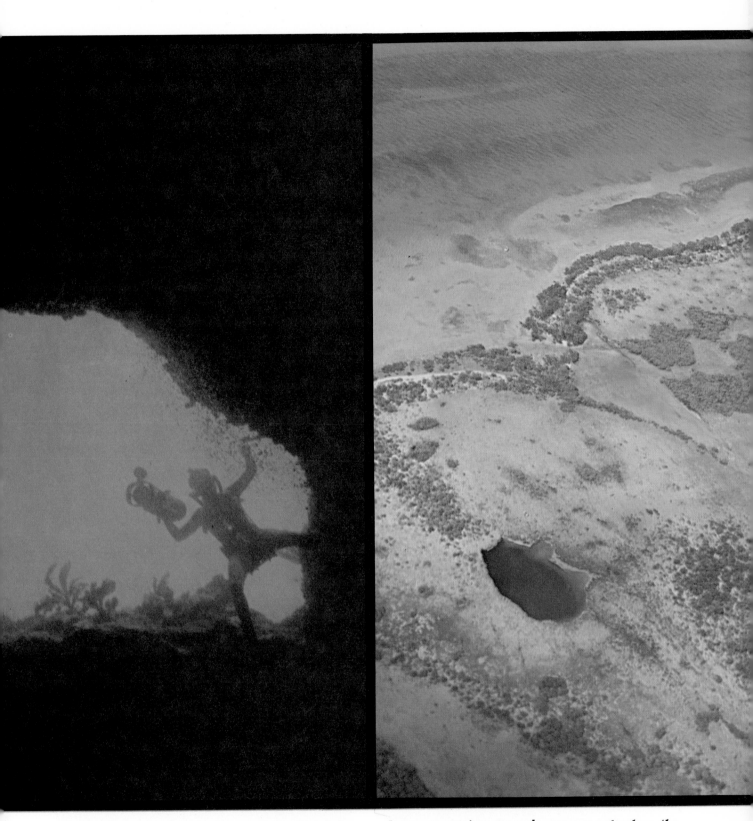

**Green dye** was poured into one of the blue holes of Andros Island in the Bahamas. Here a diver is silhouetted in the green water against an opening in a tunnel.

**Interconnecting tunnels** are proved when the green dye begins to seep from many fissures in the reef. Andros Island and its surrounding reef are formed like a giant rocky sponge.

# Chapter IV. Corals and Their Relatives

Corals are members of a group known as coelenterates or cnidarians—the flowers in a garden of sea animals. The several thousand types are rich in color and variety, from the tall sea fans and feathers to the small hydroids, from the fleshy sea anemone to the glassy jellyfish.

Among the coelenterates is a class of anthozoans containing thousands of types, including both the soft corals (alcyonaceans) and the hard corals (scleractinians). Both secrete limestone skeletons, as does the group known as hydrocorals. But only the hard or stony coral creates a limestone base that encircles the polyp and forms a colony that can constitute the framework of a reef.

The reef-building coral polyp is much like the simple sea anemone, but its functions are more complex. Both have soft hollow-cavity bodies with an opening ringed by tentacles to snatch food. Most anemones, however, are solitary, whereas the reef-building coral generates a colony with a stony skeleton that is shared by its polyps.

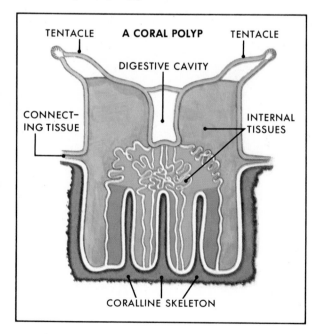

TENTACLE  **A CORAL POLYP**  TENTACLE

DIGESTIVE CAVITY

CONNECTING TISSUE

INTERNAL TISSUES

CORALLINE SKELETON

According to the environment it lives in, the same species can develop considerable variations in some of its characteristics and can grow in a variety of shapes: some species develop into bushy colonies when they live in shallow water and become flat assemblages in greater depths.

Because stony reef-building corals must have a lot of light as well as pure and warm water for their "working conditions," they are only found in a belt around the equator

---

"**Maximum growth appears at depths of less than 30 feet. The critical controlling factor in this case is illumination.**"

---

about 30° north and south, where the temperature of the sea averages approximately 70° F.; the greatest variety is found in temperatures of 77° to 84° F.

At their latitudinal limits reefs include a much smaller number of species; they also grow more slowly and are very vulnerable. Corals have salinity tolerances, too. Water turbulence and wave energy are also important controlling factors on growth. The hardness of the substrate is a factor, and so is depth. Most species stop developing at depths of more than 75 feet and maximum growth appears at depths of less than 30 feet. The critical factor in controlling this is illumination. Suspended sediments also limit coral growth. The polyps are unable to remove great quantities of particulate matter and can easily be smothered.

*Brain coral of the Caribbean* (opposite page). *Meandering grooves in which the coral polyps live side-by-side on a reef give this massive coral the appearance of a human brain.*

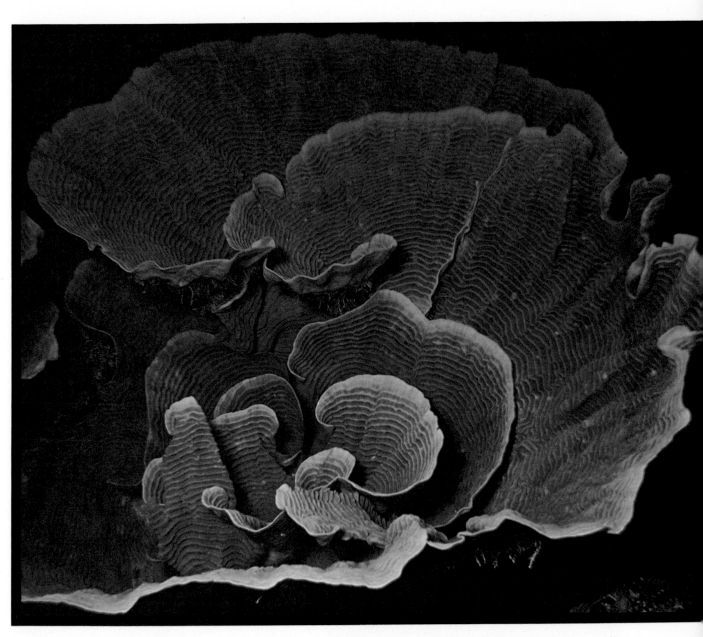

## Influence of Light

Because of the vital presence of algae in the tissues of the reef-building coral, closeness to light for photosynthesis is important. As a result, coral flourishes best near the surface where the water is clear. When water is clouded by silt stirred up by sea action, not only can mineral particles damage or clog the fragile polyp, but turbidity also hinders the penetration of light with an inhibitory effect on the coral's development.

*In deep water* (above) Pachyseris of the family *Agariciidae spreads horizontally.*

*In shallower water,* Leptoseris of the same coral family (left) shows a tighter growth pattern.

As coral grows further away from a source of light, it expands horizontally and thins out to best utilize illumination. Away from the surface, the rate of limestone production decreases. Coral types that can successfully compete for light with others in the colony reproduce faster and thus dominate an area.

## Surge of Water Governs the Shape of Corals

The movement of water has two effects on coral. It limits growth, and it governs the shape of the colonies of certain species. As a rule, encrusting, low, or heavy branched forms abound where the surge is strong. On the lee side of islands and in sheltered lagoons, colonies are more fragile and delicately branched.

The effect of water turbulence on the shape of a colony is demonstrated by *Acropora palmata.* This coral flourishes both in the breakers and in sheltered areas. In quiet water it develops into a treelike colony with a short broad stem and flat intricate ramifications extending in all directions. In very shallow or extremely rough waters, *Acropora* remains compact and encrusting.

**Delicately branched coral,** Seriatopora *(opposite), lives in deep waters or quiet lagoons. The pink color is unusual in this genus.*

**Massive elkhorn coral,** Acropora palmata *(below). Its heavy branches grow parallel to the water's surge in the turbulent shallows of the Caribbean.*

## Increased Surface for Feeding

Atoll corals are distributed according to type rather than species. Free-living, delicately branched, and foliaceous corals predominate on lagoon reefs and in sheltered pools. Stout-branched, lobate-columned, and massive corals are concentrated on shallow reef flats, toward the outer edge of the reef, and on upper seaward slopes.

Spectacular branches of coral, with tops that bloom out like umbrellas sometimes as tall as seven or eight feet or that appear as a labyrinth of coral extensions, are very common in tropical and subtropical seas of the Pacific on reef flats. They are also found in the Caribbean. Green, blue, and mauve, they have a graceful and delicate appearance in spite of their sharp branches of stone.

These large ramified corals grow at a faster rate than other more compact types, since they provide an increased surface to catch the plankton upon which the polyps feed. The thick stone lacework of *Acropora*, for example, works almost like a plankton net.

*Divers swim over beautiful lacy stone coral castles spreading in stout branches and leafy umbrellas. The wide surfaces of these coral structures give the individual polyps living there enough space to live, feed, grow, and multiply.*

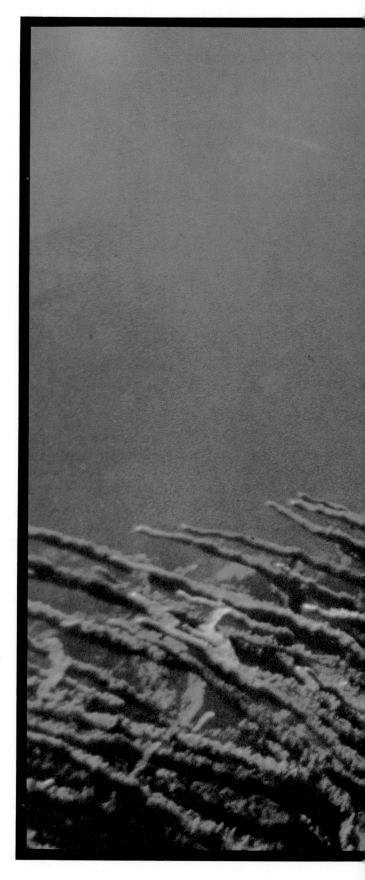

# Corals That Move

Mushroom corals, or *Fungia,* exemplify the amazing ability of coral to adapt to an incredible variety of reef conditions, especially to sandy bottoms.

Disc-shaped mushroom corals may travel on a tidal current. When they feel the need to move, they expand their tentacles like the sail of a ship.

If one is turned upside down, it is able to right itself after a long series of contortions. Its tentacles stretch out and move about as if they were searching for something. The animal soon appears to be fatigued, and its disc is shrunk. Then its tentacles gradually narrow and shrink, and suddenly the upper surface of the disc is expanded, so that one side is uplifted. The expanded body surface is immediately shrunk, and the disc becomes horizontal in position. The tentacles are again stretched out, and then they are shrunk again to prepare for the righting movement. The disc is lifted higher and higher, but each effort to turn over fails. After each failure, the coral takes a rest. Finally the body is expanded and the disc becomes nearly vertical. With great strain the coral expands a little more, and at last the disc is turned over. The coral then opens its mouth and vigorously spouts seawater. Its tentacles expand again. Then the animal shuts its mouth and hides it among the tentacles. The righting action is completed.

Sometimes *Fungia* are piled one on top of another by tidal currents. When this happens, they are able to rearrange themselves side by side. This is done by a process that *Fungia* also use to clean themselves, an expansion of their body surface. They also get rid of foreign substances or sand through ciliary movements and sometimes with the help of a secretion of mucus.

Some mushroom corals creep slowly over the bottom. They are also able to free themselves if they are covered by sediment.

Some of these corals seem to be more or less voluntarily associated with a tiny worm that lives in a cavity on its underside. The worm helps the coral by pulling it across the sand. Trails in sand have been seen that were formed when the coral righted itself with the help of a jerk of the worm.

**Solitary mushroom coral** (opposite) is able to move freely on the bottom of the sea, and right itself when it is upset.

**Fungia disc** (below) closely resembles the underside of a mushroom cap, and gives the coral its name. Disc can be five inches across.

## Fire Corals

All corals inflict wounds on the diver who carelessly brushes them, but none so painfully as the stinging, or fire, coral. One of a number of hydrocorals, it secretes massive skeletons of limestone, either in an erect branching or leaflike structure or in a low and encrusting form. The fire corals contribute heavily to the formation of many reefs, mainly in the western Atlantic and the Red Sea. They are white or of a pale fleshy or yellowish hue, and they grow to heights of two to three feet.

The individual polyps close tightly in the daytime and expand at night. Each one is very small so that fire corals look smoother than others. Their tentacles are armed with poisonous nematocysts that produce a fiery sting. Burns repeated within a few days of each other may induce severe allergies.

## Coral in Cold Climates

Coral grows slowly in cold climates for a number of reasons that are not all very clear. Daylight during the year is irregularly distributed. In winter the sun hardly shows, and the associate algae, zooxanthellae, do not get enough light to perform their metabolic processes, including calcium carbonate deposition. Most of the solitary and colonial northern coral species have adjusted to living without help of zooxanthellae.

72

Arctic waters are known to be richer in plankton than tropical waters, but this is true only for the yearly production—in winter the cold waters are empty, while tropical seas provide a fairly regular, if moderate, supply of food to the polyps. Lower temperatures also reduce metabolism. Cooler waters are farther from calcium carbonate saturation; therefore the skeletons of dead coral dissolve faster than they form. For these reasons, the formation of coral reefs is limited to warm, tropical waters.

But the fact is that some coral species such as *Dendrophyllia* or *Lubastrea* live in small groups in very cold and even in very deep water. Some large, sturdy, and branching colonies of specialized *Lophelia* develop on rocks 2000 and even 3000 feet deep. Many oceanographic dredges and fishing nets have been caught and lost in their branches.

*Solitary northern corals* (**Balanophyllia**) *shown on these pages live in the cold waters off the west coast of North America. They do not build reefs.*

# Sea Fans

Graceful gorgonians, sea fans made of a horny substance, are amazingly flexible and strong; they grow firmly attached to the sea floor and have the capability of orienting themselves according to the main surge of the water. In studying these colonies of animals, scientists have discovered some interesting points about their growth.

The distinctive feature of sea fans is their placement—most of the fully grown fans stand perpendicular to the motion of the waves. By confronting it broadside, they receive the impact of the ocean's surge more passively and simply bend back and forth. Small fans seem to face in no particular direction, but as they grow taller, their consistent tendency is to avoid the twisting that would result if they were oriented parallel to the surge. As the colony shapes itself into a fan, it acquires a larger surface to catch more planktonic food for its tiny polyps.

*Flexible sea fans* orient themselves to face the prevailing currents (above) and bend gracefully in strong currents (right). Water flowing through the horny latticework brings food that is caught by the extended polyps of the colony (opposite page).

## Sea Whips

Long stringlike sea whips are members of the gorgonian family and are related to sea fans. Like them, they have a flexible horny skeleton. They are found, among other places, in the reefs of the western Atlantic and Red Sea, growing at depths ranging from 25 to 2000 feet or more. They are highly sensitive to light and expand their polyps at night. With their thin, long, wavy, whiplike stems they resemble abandoned

*A red gorgonian* that resembles a bunch of sea whips has stalks extending from a common base. Its open white polyps show tiny branched tentacles.

electric cords, mattress springs, question marks, or canes. Not all of the gorgonians fit into distinct groups but seem to be transitional forms. Sea fans are characteristically flat, but some show a tendency toward bushiness. Many gorgonians have branches emanating from a single stalk while the sea whip is one long unbranched stalk. Some that closely resemble whips occasionally branch. It is possible that these are intermediate stages in the evolution of gorgonians between bushy forms and stringy whips.

**Whip coral.** *The whip's narrow diameter allows a large number of polyps to be accommodated with a minimum production of body material.*

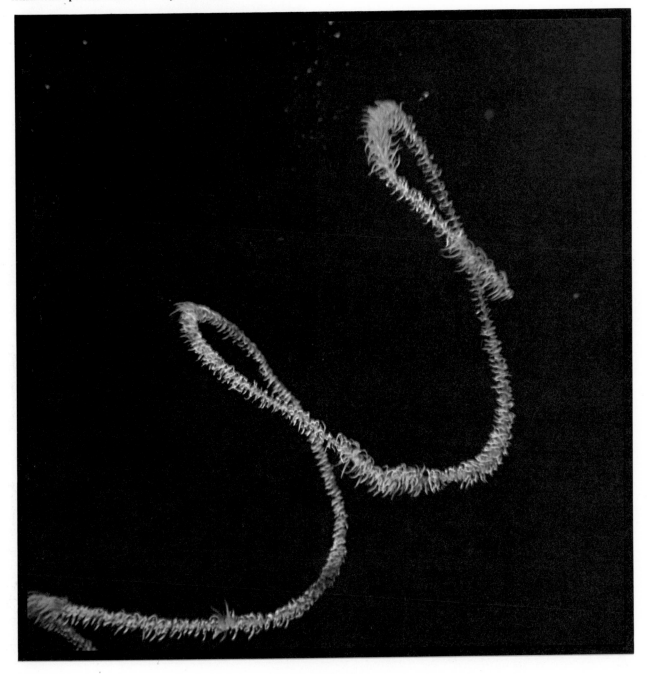

## Precious Corals

Jewelers sell precious objects made of a rare coral species that grows in deep water and has little to do with reef-building varieties. Jeweler's coral is most often red (oxblood), sometimes pink, white, or very rarely "pale flesh" (angel's skin). This prized bounty from the Mediterranean Sea was dredged with leaded wooden crosses in the time of the Phoenicians and early Greeks. In the seventeenth century free divers gathered coral off Marseilles in 50 feet of water. Since the little stone tree takes centuries to grow a foot in length, it became scarce. After 1880 helmet divers descended to 200 feet,

*A diver* (right) *discovers black coral* growing along the face of the reef. Black coral has been found as deep as 1320 feet off the coast of California.

*Red coral* of the genus Corallium (below). Avidly collected over the centuries for jewelry and objets d'art, this coral has become extremely rare.

and now aqualung divers reaching to 300 feet have practically scraped the Mediterranean bare of red coral.

Until the end of the eighteenth century physicians used coral powder as a remedy. In more recent times red coral necklaces have been worn to ward off disease and to relieve children of the pain of cutting teeth.

Brown or ochre trees, shaped like tamarisks, grow to a height of ten feet in depths of 100 to 200 feet in the Red Sea and in the Indian Ocean. They grow to more moderate sizes in other oceans and at greater depths. They are called black coral, because once their branches are cleaned of their brown polyps and polished, they produce a rich, deep-black precious material.

Black coral trees are made of a horny substance and look like gorgonians. They are not. They are known as antipatharians and are related to true corals and anemones.

In order to grow well, black coral needs a sea substrate clear of sediment. It is found in caves or beneath overhangs where the light is subdued. Unlike other coelenterates, black coral thrives in strong deep-ocean currents, away from surface light. Recently observers in research submarines have found beds of black coral in waters of the Coronado escarpment of California at a depth of 1320 feet and in the San Diego Trough.

Black coral takes a superb polish and has long been a treasured raw material for jewelry and objets d'art. It is considered as sacred material in Islamic countries, where it is turned into expensive prayer beads. In China, Japan, and Indian Ocean countries it provides the material for charms to ward off bad luck and disease.

*Plumelike branches* of black coral that thrives in swift currents provide shelter or even a home for many small fish, crabs, sponges, and anemones.

# Flower Animals

As numerous as dandelions in a field are anemones in the sea. About 1000 species have been found in both Atlantic and Pacific. Their color and form vary so much that they have been given many appropriate names: brown sea anemone, powderpuff, white plume, dahlia, and so on.

The treelike alcyonaceans are also anthozoans, or flower animals. They often have large, soft polyps, and some pulse rhythmically, making it obvious that they are alive. If kept in the dark, soft corals show signs of starvation within two weeks. Evidently light is more essential to them than food, and the polyps' association with the algae that live in their tissues is vital to them. Because of the zooxanthellae, they have become dependent on light to function and survive.

*Sea anemone* (below) and yellow tube sponges of the Caribbean. Tiny bumps on the anemone's tentacles are stinging cells.

*An alcyonacean* (right) is a big rubbery soft coral common to the Pacific. The treelike body is stiffened by scattered spicules.

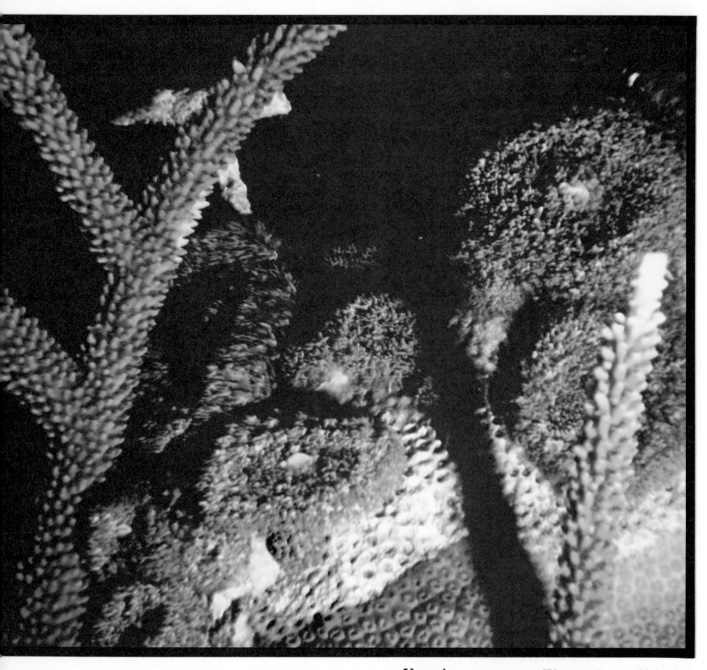

***Usurping anemones.*** *The coral on which these anemones have settled may have been weakened in advance by grazing fish or molluscs.*

## The Fight for Space and Light

Corals sometimes kill each other to find space and light in which to live. This rarely happens with corals of the same species but often involves closely related types.

It is generally in the buffer zone of the coral reef, where overpopulation is most likely to occur, that cannibalism is used to solve the problem. No such fratricides have been observed among corals where there is room and light enough for all.

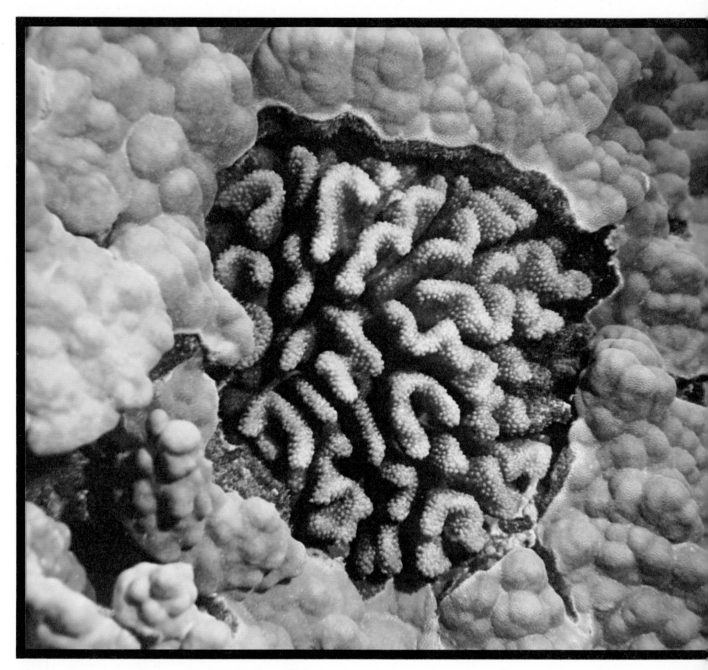

*Surrounded, the **coral** at the center of this photograph appears to be expanding its territory aggressively at the expense of its surrounding relatives.*

Aggressive acts are set off by physical contact. The stronger "aggressor" polyp attacks by extending mesenterial filaments over the weaker coral, literally digesting its prey outside of its own cavity with the filaments. Thus such battles, although they are initiated as struggles for space and light, revert to the basic, simple feeding response set off by contact between creatures of different species. The victim in the competition becomes a source of food the same as any other unrelated prey would be.

# Chapter V. The Coral Farmer

Zooxanthellae are always found in reef-building corals, where their concentration varies among the different species. Their symbiotic relation with their hosts has been an area of particular interest to scientists. It was not always understood whether it was the plant or the animal which was "farming" or "exploiting" the other. In any case, the zooxanthellae aid the coral's nutrition, recycle its waste products, and help it secrete its skeleton.

---

**"Corals might have taken to farming symbiotic algae to gain advantage of the sun's energy."**

---

One question that has been much debated is the extent to which the algae actually contribute to the nutrition of coral. It was once believed that the purpose of the algae was to provide plant food. But corals have been shown to be strict carnivores. A crucial experiment that helped confirm this was one in which reef corals were kept in the dark for 228 days and fed on zooplankton. They stayed in good health. Then they were kept in the light but without zooplankton for only nine days. At the end of this time they began to show signs of starvation. Three days later they died. It was thus clear that they could not subsist on the plants or on photosynthetic products of the plants.

The algae release to the coral organic compounds fixed through photosynthesis. Where food is limited, the algae thereby represent a reservoir of reduced organic compounds. Where food is abundant, corals could be expected to live perfectly well without the services of zooxanthellae.

Zooxanthellae, like all plants, produce oxygen as a by-product of photosynthesis. This is presumably available to the coral, but its physiological importance to the animal is still uncertain.

Zooxanthellae also function like a kidney to the coral, removing toxic waste from its linings. It also appears that ammonia excreted by the coral is returned in the form of amino acid and that in this way nitrogen is saved and can be used again.

Phosphorus is another ingredient of coral wastes. It is taken up by the zooxanthellae to satisfy their own needs, then recycled and returned to the animal. The symbiotic algae also conserve nutrients by accumulating those products and returning them to the coral in a useful form.

We also know that corals with symbiotic algae calcify many times faster in light than in darkness. It has been suggested that this is because the algae, as a normal part of their nutritional functions, remove phosphates from the environment, thus favoring the crystalization of calcium carbonate; or that the removal of carbon dioxide through photosynthesis enhances the precipitation of calcium carbonate. A third hypothesis is that organic products of photosynthesis permit faster calcification.

Paul A. Zahl, senior scientist of the National Geographic Society has asked: "Have the animals of the reef, unable to do what the plant does, taken to symbiosis to get as close as possible to the source of trapped energy from the sun?"

*Green coral colony* (Monastrea annularis). *A tiny goby pauses on the velvety green of the corals' surface where it finds some protection from predators. Some of the coral polyps are extended, others of the colony are tightly closed.*

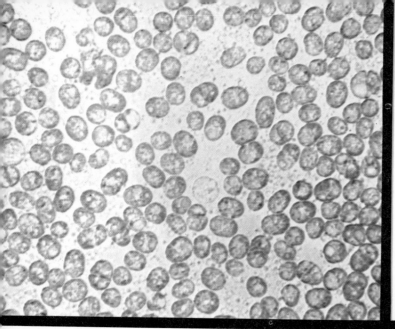

*Zooxanthellae. Microscopic one-celled plants, greatly magnified in the picture above, give a green cast to many corals they live with.*

likewise important to the algae, which make use of the polyp's carbon dioxide, ammonia, and phosphates. Because zooxanthellae are plants, they seek the light of the sun for photosynthesis—the production of organic substances, chiefly sugars, from carbon dioxide and water with the aid of chlorophyll. It is within the upper 200 feet of the oceans that photosynthesis can take place. And yet zooxanthellae have been found living with anemones at a depth of 1250 feet off Antarctica and with anemones and corals at a depth of 1920 feet off Key Largo, Florida. This has

## Nutrition and the Coral Algae

Coral polyps are carnivores and the majority of them feed at night. Their diet consists primarily of planktonic animals. They trap and transport this food to their mouths with tentacles that are laden with stinging structures (nematocysts). Food that is too small to be grasped by the tentacles is trapped on a sticky, mucus-laden area between the polyps of a coral colony. This food is transported to the mouths in moving sheets of mucus. In this way, a colony with its polyps expanded represents an enormous feeding surface in relation to the bulk of its living tissue. It was recently proved that at least one kind of coral eats fish. Photographs showed that the tentacles of a coral polyp can sting a tiny damselfish, move it to the mouth, and ingest it.

It is known that zooxanthellae, the one-celled algae that live between or within the cells of the coral polyp, are of great ecological importance to the coral's growth, but it is not yet fully understood how they contribute to the coral's nutrition. The coral is

86

led to a suggestion that at the expense of the host animal, these deep-dwelling zooxanthellae live by obtaining food only from organic material. But what does the polyp obtain from the algae that might aid its nutrition? One difficulty in answering the question is that the total nutritional requirement of a coral is still unknown. There is evidence, however, that corals acquire soluble organic carbon compounds for their nutrition from the zooxanthellae. We know that a translocation of organic material from the algae to the metabolic pathways of the coral does occur, but the significance of that movement in terms of quantity is not yet known.

We do know that corals with larger polyps are more efficient at capturing zooplankton and that they have lower rates of metabolism. They are therefore less dependent on zooxanthellae for their supply of carbon than are corals with small polyps.

**Brain coral** *of the genus* Colpophyllia. *Several polyp mouths can be seen in the valleys and crevices between the high coral walls.*

# Calcification of Corals

We have already seen that zooxanthellae, the symbiotic algae of corals, play a fundamental role in the growth of the coral's skeleton—with the help of the light's energy, the presence of zooxanthellae significantly accelerates the rate of calcification. Like other plants, under the influence of light the zooxanthellae consume carbon dioxide during photosynthesis, and the carbon dioxide functions in the acceleration of calcium deposition. This hypothesis is inconsistent, however, with the observation that polyps at the apex of branching corals calcify much faster than lateral ones even though the apical polyps have relatively few zooxanthellae. It might also be asked how carbonate is supplied for calcification if the algae remove carbon dioxide from the environment.

Two alternatives to the "carbon dioxide removal" hypothesis have been proposed. One is that the transport of metabolic products released by the algae toward the apex of a branch (for which there is some evidence) might explain the rapid calcification of the apical polyps. The other is that the algae remove phosphates, which are also inhibitors of calcification, from the calcifying milieu. This may explain why corals in darkness with algae calcify significantly faster than those in the dark without algae. Increased removal of phosphate in the light may thus explain how light accelerates calcification.

*White tips of coral* (opposite page) *are evidence that growth in these areas is faster than it is elsewhere in the colony. The location of this rapid growth is puzzling.*

*Growth paradox. Consistently faster growth occurs at the tips of coral as seen below; but it is here that the symbiotic algae so significant to coral calcification are least abundant.*

# The Growth of a Coral

Just before and during a new moon, a coral polyp sets free hundreds of larvae, called planulae, from a fertilized egg. For five to nine days these larvae half swim—with the help of hundreds of tiny hairs that cover their bodies—and half drift on the currents of the sea. The larvae from mushroom coral only drift for two or three days, others might drift for more than three weeks.

Almost immediately after the larva is released, the beneficent zooxanthellae algae distribute themselves throughout the larva, giving a brown hue to its translucent body.

The larva's journey is not without its perils. Muddy water can choke it. Protozoa by the hundreds can attack it and digest it.

If it survives, the larva thickens and settles onto some smooth surface. Almost at once it begins to secrete its limestone skeleton, creating a shelter and cementing itself to its homesite. The larva is transformed into a polyp. Its center sprouts into a hollow pedestal crowned by tentacles. Some corals de-

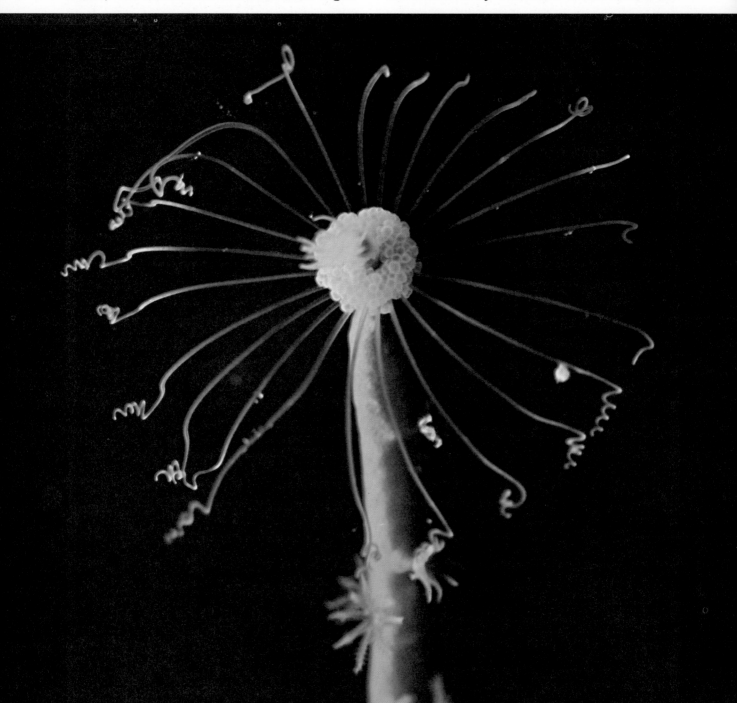

velop no further. They are isolated and are joined to neighboring polyps only by thin appendages at the base. With the proper environment, however, a single hermatypic coral polyp can be the beginning of a whole new colony of individuals.

One way in which coral polyps reproduce is by asexual budding. New polyps bud forth from the cells of this connective tissue and immediately begin to secrete their own stony cups. Polyps also reproduce asexually by fission, putting forth more tentacles and fleshy partitions until they split from the mouth down, and there are two where there was one. All this time the skeleton of each polyp becomes more rigid and complicated.

When a polyp is mature, spermatozoa and eggs begin to develop within its mesentery. When their maturation has reached a certain point, spermatozoa flow through the mouth of the polyp and float in the sea until they reach other polyps. The other polyps draw them inside with fanning movements of their tentacles. The spermatozoa then fertilize the eggs that have remained attached to the inner linings of the host polyps.

Soon after fertilization, the mesentery releases the eggs. Like the spermatozoa, they flow through the mouth of the polyp. The larvae develop into hollow spheres of cells perhaps the size of a pinhead. It takes from a few hours to a few days for the pear-shaped larvae to settle and become polyps.

*Corals reproduce both sexually and asexually. A cross section of a polyp (top) with fertilized eggs that will hatch through its mouth demonstrates a stage in sexual reproduction. Budding (center) and fission (bottom) are asexual phases.*

*Hydroid (opposite page). Like their coral relatives, hydroids reproduce both asexually (by budding) and sexually. Long tentacles on the polyps capture food. Reproductive polyps break off as free-swimming medusas and settle down away from parent.*

## The Total Ecosystem

Just as scientists consider the coral reef to be a macrocommunity, so the individual polyp itself is considered to be a microcommunity. It shows its self-sufficiency with its symbiotic algae acting as gardens to provide the host with the nutrients it needs. On land such an association of plant and animal is external; a cow grazes in a pasture and a wolf eats a lamb that has fed on grass. In any case the raw food material has been generated by plants through photosynthesis. In the case of coral, the necessary plant lives inside the

polyp. It is (very roughly) as if a cow had its own prairie inside its body.

The surfaces of the coral head provide for its own maintenance and growth. Tiny animals, borers and foulers, are soon attracted to each type of coral colony, and the result is an extremely complex struggling jungle personalized with each coral head. The reef is a mosaic of these small habitats.

Each coral head helps recycle species. Its remains become the foundation on which other live coral polyps will build.

# Chapter VI. A Coral City

Coral reefs have been called "nature's cities." The comparison with large man-made towns is striking. We will review in the next pages how the major functions of the coral city are performed. The harmony of its existence today is the result of many millions of years of struggles, failures, and successes.

Reefs have enormous rates of organic production, rivaling the efficiency of man's best agriculture. This is particularly remarkable when we measure it against the relatively low average productivity of the oceans. The open-ocean waters, from the very surface to about 4000 feet of depth, produce altogether the bulk of the primary and secondary plant and animal marine life. But this complex broth is thin; it looks almost empty to the eye of a diver except in rare agglomerations.

The bottom of the ocean, though never a complete desert, has also a majority of bare provinces. When the bottom is well populated, it generally offers communities (benthic biomasses) changing gently from one area to another but often relatively simple. It seems that when a bottom ecosystem is locally successful, it manages to eliminate intruding species or to repel them to the next ecosystem which is often quite close but different in composition.

This is why the fantastic jungle-cities displayed by coral reefs are so remarkable. For miles, or sometimes for hundreds of miles, tens of thousands of species belonging to practically all known phyla coexist in a competitive but successful balance.

All living ecosystems can be sketched as pyramids, with algae and other plants at the base and with the most advanced carnivores at the top. If the base is narrow, the system is fragile, and the extinction of a few species may bring about the collapse of the pyramid. This is the case in all polar waters, which are poor in variety but rich in quantity of living organisms.

In the case of the coral cities that we are studying in this book, the pyramid is amazingly broad, much greater than in the thickest jungles on land. If man did not interfere, the system would prove to be the richest, the most varied, and in case of localized disasters, the most resistant on earth.

---

*"The coral cities have a rich and varied population with their producers, their consumers, and their waste recyclers."*

---

A simplified way of thinking about a reef is in terms of producers, consumers, and waste recyclers. The primary producers are plants that must live where sunlight is available. The plant-eating consumers release simple compounds of carbon, nitrogen, phosphorus, and other elements. Regenerated nutrients are returned to the plants, completing the cycle of matter. There are "leftovers," deposited in the fissures, in the lagoon, and along the slopes, and these are in turn eaten by sea cucumbers, worms, snails, and many other animals, some of them microscopic in size. The nonnutritive sand or mud, mostly calcium carbonate, is pushed by waves to build beaches or is piled up in large dunes deep in the ocean.

*Burghers of the reef. Prominent citizens of the coral city are grunts, seen here schooling in and around branches of elkhorn coral. More distinctly marked porkfish in the foreground are also members of the grunt family, Pomadasyidae.*

# Inhabitants in a Crowded City

The coral reef "commune" is a very crowded city with its skyscrapers, its business center, its residential sections, its suburbs, and its slums. It is regarded as a structure in a steady-state balance of growth and decay only modified by slow changes in the sea level or in climate. All its members adapt to and benefit from each other in the relatively small area of the reef city.

The reef has its producers, consumers, and waste recyclers, with each member—even the bacteria that process sediments—playing a vital part. Some members bore passageways in the reef that keep it ventilated; predators help maintain a healthy balance of population; filtering animals keep the water clean, which helps light trigger photo-synthesis. The community produces its own nitrogen and recycles phosphorus. The reef reaches down with its nooks and crannies, corridors and crevices. They are used for shelter, for hideaways from aggressors, and for reproduction.

Small fish like gobies or blennies live among the very polyps of coral. Some never leave their tubular sponge homes; some fight to defend their shelter, while others are more hospitable to intruders. Because the structure of a reef provides only a given number of sheltering crevices, a natural self-limitation of fish life occurs on the reef.

*Squirrelfish. Resting in the crevice of a wall encrusted with purple coralline algae is a big-eyed nocturnal squirrelfish, right. At night (below) it ventures out to forage around the base of pillar corals (Dendrogyra cylindrus).*

# Cementers of the Coral Reef

Each reef is a gigantic mountain of dead organic matter crowned by a shallow but exuberant layer of living creatures. It would easily crumble against the pounding of the sea if it were not held firmly together by some kind of cement. This cement is coralline algae.

At the seaward edge of the windward reefs there occurs a rich growth of algae. The globular, spongelike, red calcareous algae flourish in the teeth of thundering breakers.

Other organisms are almost totally excluded from this most turbulent of reef zones.

The entire seaward side of reefs is surrounded by a skin of massive corals and encrusting algae, which serves to contain the less rigid skeletons and sediments within the reef. Such a protective belt does not prevent the reef from being remarkably porous.

*Broken coral branch* (*opposite page*) *will become part of the debris* (*below*), *which will eventually be stabilized by dead plants, animal shells, and seaweed to become the base for a new living layer.*

## Farming on the Rooftops of the Reef

Since coral reefs grow up to the surface like houses, with bright varieties of coral and an algal ridge showing near or even above the waterline to utilize the maximum light, it has been said that coral "farms on its rooftops." From this observation some architects have suggested that city dwellers, cut off from the light, might emulate the coral reef

by cultivating their rooftops in areas especially constructed above their homes. In addition to a greater utilization of "airspace," which is already a factor in architectural design of buildings, there are other advantages to rooftop agriculture. Greenery is not

**Table coral,** Acropora, *grows on a thick stalk and spreads out to catch the maximum sunlight for algae it farms on its upper surface.*

only aesthetically pleasing, but it is an effective antipollutant, and locally grown vegetables would benefit city dwellers.

## The Lagoon

A lagoon, with its relatively high concentration of plankton, other nutrients, and waste material and with the turbidity of its water, forms a separate and distinct ecosystem from the outer reefs.

The lagoon floor generally lies at depths between 75 and 150 feet and rarely at as much as or more than 200 feet. Water circulation is sufficient in most places to provide near-optimum conditions for the growth of a great

*Shallow knoll is made up of the coral* Siderastrea siderea, *sponges, tube worms, brown algae, and sargassum weed.*

number of corals and other reef organisms except the red calcareous algae, which requires a higher rate of water renewal.

There is a steady inflow of water through channels on the reef surface into the lagoon on the windward side. On the lee side water transport is largely influenced by the tides. Wind drives the surface water in a lagoon in

102

a steady flow from seaside to lee. The greater part of this water cannot escape, so it is driven below to return upwind as a bottom current. Measurements of water flow into and out of Bikini lagoon show that the total volume of water is completely replaced every 39 days during the trade wind season and in about double this time during the summer. The Aldabra lagoon in the Indian Ocean is much shallower and has a water replacement rate of a few days.

The shallows of the lagoon are littered with debris of coral, of green thalassia seagrasses, and of the thin, branching segments of a cactuslike growth, halimeda algae. The lagoon is often cloudy because it is laden with the residue of animal life on the reef. Bacteria precipitate the calcium from this residue, and thus the bottom of the lagoon is covered with calcium carbonate.

If the secluded waters of the lagoon are examined closely, they prove to be a very lively and colorful place indeed.

*Seaweed,* a green algae, grows in the lagoon shallows using the ample sunlight for photosynthesis.

## Primary Grazers on Algae

To describe marine animals that depend on algae for food, scientists say they "graze" on the reef. This conjures up an image of a reef as a farmland with stony pastures.

As early as 1963 oceanauts from Conshelf II in the Red Sea and more recently aquanauts from Tektite II in Lameshur Bay of the Virgin Islands observed in detail the habits of herbivorous fish as they tore various blue green, red, green, and brown algae from coral rock and other calcareous surfaces. Some fish only scrape the surface with their mouths, but others bite little chunks of the reef and crush it in their mouths or in their throats, retain the bits of algae and spit out or reject the crushed limestone in little white clouds before they take another bite. Divers can distinctly hear the noises made by all this tearing, chewing, and crushing. Sea urchins were also seen feeding on thalassia seagrasses in shallow water during the day, while molluscs and crabs worked over

the inner reef during the day and disappeared into the outer reef at night. Sixty species of hard-mouthed fishes representing three families depend entirely on algae for food. They constitute an intermediate trophic level in the sea's food chain.

In a heavily grazed algal zone, bare whitish patches showing the intense food-seeking activity of these animals can readily be seen. Curiously, wherever herbivorous fish graze, greater varieties of algae grow.

There are other organisms on the reef—such as the giant clam (*Tridacna*)—that are able to feed on the very algae they host, gathered from the mantle of the clam and digested. While the clam probably is capable of finding other nutrients in the water that it circulates, it is a unique case in the reef with its self-dependent, "closed" system.

***Herbivorous parrotfish** bites off a piece of rock-hard coral to obtain the nutrients it holds. These fish have fused teeth in a cutting beak and grinding teeth in their throats.*

# Space Sharing on a Reef

Reefs are very crowded. A good way to take maximum advantage of available space is to organize day and night shifts. Herbivores graze by day. Many carnivores feed by night.

Carnivores can be seen hovering during the day; some of them waiting for the night to hunt, others staging feeding frenzies at dawn and dusk, all of them ready to take advantage of an easy meal at any time an opportunity presents itself. Sometimes a diurnal fish can be attracted by a bright moon to come out at night, putting itself at the mercy of the nocturnal fish.

At night the greatest activity is out over the open sand flats when most of the daytime species have sought shelter in deep holes, crevices, and overhang areas of the reef.

Many molluscs and crabs show day-night cycles as well, in which they move to the outer reef surfaces by night and retreat to the inner spaces by day. In effect, they are the harvesters of the daily production of plants of the reef.

Space sharing on a reef also depends largely on how the balance of population is maintained. Predation is the main factor of population control. The reef undoubtedly has its emigrants too—fish that leave the reef to forage elsewhere—but their number is small. Predation is most intense soon after reproduction: inexperienced fry make many mistakes, and swimming into the wrong habitat makes them easy prey for larger carnivores.

*Nocturnal foragers* (*opposite page*). *During daylight these French grunts school for safety; their coloring so closely matches the sun-dappled background that individuals are difficult to see.*

*Basket stars* (*below*) *become active at night when they spread their lacy tentacled arms to catch food. Different feeding shifts greatly increase the efficiency of the reef community.*

# Corals in the Food Chain

As far as is known, corals are carnivores rather than herbivores. They feed on zooplankton, especially at night. Some scientists believe that the coral also receives nutrition from the zooxanthellae.

Food chains build up like a pyramid, the quantity in weight of lower animals being much larger than that of the higher ones; each step up represents a smaller mass of individuals. From one step to the next, the ratio varies from three-to-one, in the case of excellent efficiency, to from ten- or twelve-to-one in most cases. Accurate measurements are difficult to make, but some indicate that the coral polyps ingest in their lifetime hardly more zooplankton than their own weight. There is a theory that the symbiotic algae, feeding from organic components, may contribute to the nutrition of the coral, but the accuracy of the experiments is still in doubt. In any case, an argument in favor of the "coral-algae" association may be to consider it a high-efficiency system involving the production, utilization, and recycling of nutritive material.

Certain fish feed directly on corals, by browsing on the polyps, by grazing on living coral heads, and by feeding on branching coral tips. Some polychaete worms have been observed feeding on coral polyps and so has a species of crab. The most alarming case of feeding on coral polyps is that of the voracious starfish, crown-of-thorns.

Most corals secrete copious quantities of mucus, and it is believed that many of the deposit feeders utilize this source of food.

*Treelike coral formations* (right) *grow up and outward for greater exposure to light.*

*This* **coral head** (below) *is tinged green from the microscopic plants that grow within its tissue.*

# Feeding Interdependence

It is primarily through food web interactions that the coral reef achieves its biological stability. All individuals are in some direct or indirect way dependent on all others in a complex feeding chain.

Here every conceivable strategy of survival has been developed, and every possible feeding method is in actual operation. Obviously, the herbivores are commonly the most abundant marine animals of any size in a reef community, and all of them are susceptible to falling victim to more powerful carnivores.

Small fish such as herrings, sardines, and anchovies literally take their choice from the

**Barracuda** (left) is a large predator, near the top of the ocean's food pyramid.

**Carnivores** of the reefs (below), an octopus and a Nassau grouper, keep wary eyes on each other.

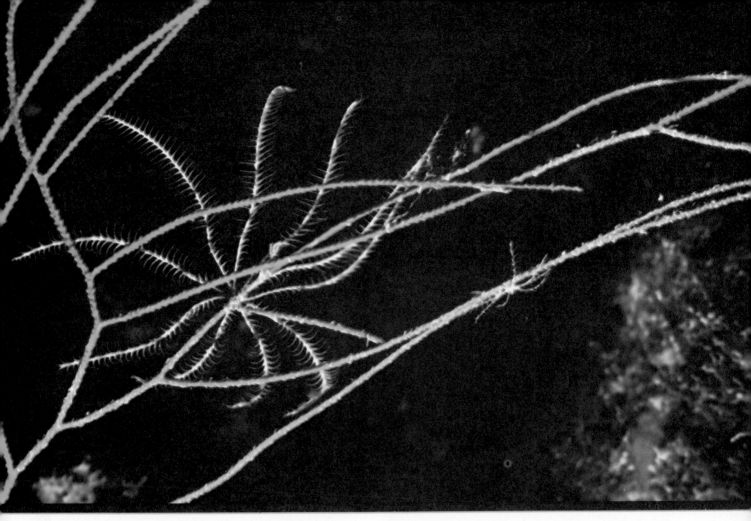

overwhelming quantity of tiny planktonic animals, picking them one by one. On the contrary, the huge manta ray is equipped with strainers to extract most of the plankton from the water that passes through its gills, and it swallows the aggregate food without discrimination. Fast-swimming tropical predators—tunas, jacks, sharks, squids, or barracudas—prey on the smaller vegetarians or plankton eaters. These predators are equipped with well-developed eyes and efficient jaws. Among the bottom-inhabiting animals, the sea stars are voracious feeders, preying on clams, which they exhaust and force open by applying continual traction on the two halves. Predacious snails, noteworthy for their long siphons, are able to drill holes into bivalves and other molluscs, insert their siphons, and eat the soft contents. Octopods are also capable of drilling holes in shellfish.

*Filter-feeding feather stars wait on the branches of a gorgonian for waterborne food.*

The filter feeders capture tiny particles suspended in the water. Deposit feeders, which include sea cucumbers, worms, snails, and many other animals, crawl over the internal surfaces of the reef, eating the organic leftovers and microbes contained in the sediments. But these waste recyclers are by no means the end-of-the-line: such large fish as the venus tuskfish fan the bottom silt away with their fins to expose worms and other deposit feeders and eat them, which pretty well loops-the-loop in what adds up to an unbelievably complex web of interaction.

In the inner reef crevices there are predators as well. With ease, the mantis shrimp can slash apart crabs in a few seconds. Hidden groupers can use their gill plates as pumps to suck in unwary passersby.

111

## A Regular on the Reef

The trumpetfish moves languidly and alone in the waters of the reef like a drifting twig. It seems to have no place to go and nothing to do. Lazy, one might conclude. But it is not lazy at all. It is merely hiding. If a predator were to discover that it was not a twig but a tasty trumpetfish, it would swish its tail like any other fish and hie to the closest crevice.

When the coast was clear and it came out again, it might go stand on its head among the sea whips. Here it would be safe. For a while at least. What big fish could possibly take it for anything but one of the tasteless branches of the gorgonian?

Of course, it wouldn't just be hiding. It would be hunting as well. If a smaller fish or a shrimp wandered beneath, it might soon vanish in trumpetfish's telescopic snout.

If a grouper came along or another big fish, trumpetfish might tag along for a hunt farther afield. It would swim as close as it could to the other fish without annoying it unduly so that it would appear that the two were

only one. The big fish might not appreciate the company, but it would probably endure it. And anyway, it wouldn't be for long before trumpetfish would find what it was looking for and dart down to gulp it.

These extraordinary creatures, which would be the last to blow their own horn, are not at all uncommon in the waters of the reef. Perhaps being not so lethargic acrobats has made them one of the fittest of the fishes.

*Making the rounds* of the reef, a slender trumpetfish pauses frequently to hide itself.

*Cleaning station.* Little wrasses (above) sometimes set up regular stations for client fish that come
to the reef from the open sea to be cleaned.

## The Cleaning Service

Apes and men use their hands to clean and
take care of their own bodies. Many animals,
like panthers or dogs, use their tongues and
teeth to tidy almost any part of themselves.
With no hands and stiff bodies, fish cannot
clean themselves except by rubbing against
rocks or in the sand, and they are helpless
victims of all sorts of parasites. Many of
them depend largely for their health, and in
some instances for their survival, on finding
a way to get rid of their parasites.

One of the services that the reef community
offers to some of its inhabitants is precision
cleaning. Some small fish and some crusta-

*Hogfish* (opposite, bottom) receives the attention of several neon gobies who rid the bigger fish of parasites and diseased tissue.

*Yellow-and-purple beau gregory* (above) picks at the skin of a Bermuda chub. Cleaners often swim into the gill slits and mouths of their clients.

ceans obtain the major part of their food supply by picking parasites and bits of decaying material from the bodies and from the mouths of "clients." The "cleaners" are thus key organisms in the health of the animals of the community.

Certain brightly colored tropical shrimp, associated with anemones, have been observed picking and eating parasites and injured tissue from a variety of fish. These shrimp remain stationed on or near an anemone, attracting fish by their brilliant colors, conspicuous patterns, and gesticulation. There can be a strong degree of competition among those shrimp for a prominent position on the anemone. The largest individuals

station themselves on the tentacles or near the oral disc of the anemone, while the smaller ones wait on the sand around the anemone or on surrounding vantage points. When a client fish approaches, the larger shrimp attract its attention by dancing and waving their antennae. When the fish has been successfully solicited, all the shrimp join in the cleaning operation.

Some of the cleaning stations probably have such a reputation that their customers (including open-ocean cruisers) have to wait in line for their turn to be serviced.

Some fish pretend to be cleaners but are not. They employ their mimicry to get within range of a prey much larger than they are.

# Reef Sewage Treatment

The life processes on a reef produce two sorts of waste: reusable organic material, including bodies of dead animals, mucus, and fecal matter, which must be recycled because it is rare and precious, and inorganic sand, which must be eliminated. The coral community is organized to perform these two functions. Dead animals are eaten by scavengers; crabs, shrimp, and lobsters are the caretakers of the coral city. This category of animal feeds on any corpse, including those belonging to the same species—an interesting case of scavenging cannibalism! As a result, practically all organic tissues are transformed into fecal matter, the main exception being

a variable fraction of the population of bottom crawlers and burrowers.

The resulting waste settles in the cracks, holes, and crevices of the reef, on the bottom of the lagoon, and to a lesser extent outward along the steep slopes where the material is practically lost for the citizens of the reef. The second stage of waste treatment is per-

*Colonial sea squirts* (above) are the most primitive animals of the phylum Chordata.

*Sacklike bodies* of sea squirts (below, left) have one opening to draw in water, another to expel it.

*Ascidians* (below, right) encrust rocks much as sponges do. In larval stages they are free swimming.

*Crimson tunicate* (opposite page) has a single opening surrounded with a fringe of cilia.

formed by the deposit feeders (worms, snails, sea cucumbers, and so on), crawling over or under the surfaces of the mud, treating enormous quantities of sediment from which they extract organic debris and fecal matter. This service keeps sediments from blocking internal passageways, provides ventilation, and resuspends particles in the water to be reprocessed by filter feeders.

The third and final stage is performed by bacteria that teem in sediments. In addition, all calcareous surfaces harbor microbes and fungi which often penetrate skeletons to considerable depths. Bacteria return any organic matter to its fundamental components, basically nitrates and phosphates. Bacteria are oxygen consumers and would asphyxiate the reef if their population grew unchecked. But they are also consumed in great quantities by larger creatures.

Even fish can be part of the process. The sand tilefish construct their burrows with coral fragments that they carry in their mouths, much as a bird builds its nest with twigs. They are capable of moving great masses of coral rock.

# The Water Conditioners

The reef contains its own water-conditioning system. The internal spaces of the reef function as a trickling filter. In addition, many filter feeders, such as sponges and clams, capture tiny particles suspended in the water.

Sponges have millions of microscopic pores that pierce their surface. Through these pores the sponge sucks the water that contains microscopic plants and animals that it feeds on. A steady stream passes through its body and out the large vents at the top.

The volume of water that a sponge can pass through its body is astounding. One estimate set the rate for a Bahamian wool sponge at about two quarts a minute.

All of the filter feeders induce feeding currents that help to ventilate the internal reef spaces, thus perfecting a very elaborate water-conditioning system.

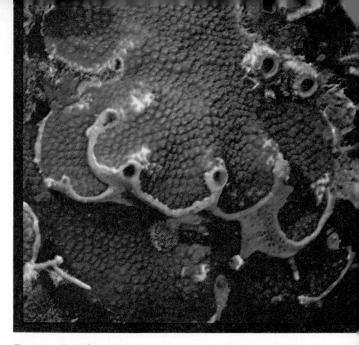

*Brown border of sponge* edges coral formation (above) and clearly influences both the shape and size of the coral colony.

*A living haystack* (opposite page) is Hydnophora coral. The orange sponge beneath it constantly pumps water through its numerous pores.

*Green coral and brown sponge* grow together on the reef (below). Little orange corals beneath the sponge are not primary reef builders.

# Waste Recyclers

Bacteria, worms, and holothurians are among the recyclers of materials on a reef. The bacteria return waste to the system as nutrients. The burrowing animals sift out from the substrate material that would otherwise be lost. They act like garden worms turning over the sediments. These organisms allow the reef to remain an almost closed system by eliminating the accumulation of waste. The only waste that remains on a reef is calcium carbonate, which is used to build the coral city itself, the excess being eliminated as coarse sand along the slopes of the reef, which it thickens and reinforces.

Bacteria are found everywhere in the oceans, but in especially high concentrations in reef environments. Their metabolism is similar to that of animals; neither can make its own food, as green plants do, but must extract the energy and organic compounds necessary for growth from preformed organic materials. As a general rule, they oxidize about 70 percent of the organic matter used and convert about 30 percent of it into new bacteria, a very high efficiency ratio.

The sausage-shaped sea cucumbers crawl very slowly on the muddy bottom. They are about a foot in length and two inches in diameter. At one end of the sea cucumber a set of tentacles slowly takes in food. These soft organs around the mouth shovel the surface mud into the digestive tract, and in this way the animal obtains the nourishment it needs from a great variety of microscopic life, especially diatoms. The gritty residue is expelled from the other end. The sea cucumber can pass as much as 200 pounds of sand through its body in a year.

*Sea cucumbers, whether living in temperate or tropical seas, perform the essential service of turning over and recycling sediments.*

120

## Visitors to Coral Reefs

Open-sea fish accustomed to traveling hundreds of miles in the blue expanses of the ocean find the reef waters an inviting place to gather and breed, to forage for food, or to use the facilities of a cleaning station. Pregnant hammerhead sharks seek shelter in a reef bay to give birth. The reef provides abundant food for the young sharks. Tuna and jacks from time to time linger around before returning to the open sea. Even whales sometimes venture in.

With the coming of autumn, thousands of birds leave Canada, Alaska, Siberia, and other North Pacific lands to fly to milder climates. Some of them stop on the atolls of the South Pacific. One of the migratory birds commonly seen on atolls is the Pacific golden plover. Other plovers, curlews, sandpipers, and phalaropes also visit to rest on their annual migrations.

Great marine turtles are occasional visitors to coral reefs. Green, loggerhead, or hawksbill turtles gather in shallow waters and the females lumber onto shore to lay eggs.

**The pompano** (above) is one of the many open-sea fish that can find a visit to a reef a profitable venture.

**Striped sea bass** (below) that range the Atlantic coast are frequent visitors to reefs of the Caribbean.

# Chapter VII. Coral's Predators

Until fairly recently corals were thought to be virtually immune to parasites and predators. Their stinging nematocysts were assumed to be an adequate defense. But now we know that a surprising number of animals feed on coral. They include fish, asteroids, crustaceans, polychaetes, and gastropods. New ones are still being discovered. Recently the comb-tooth blenny was added to the list of coral's predators.

Many herbivorous fish graze on the algae of the reef. Some carnivorous fish feed directly on the corals, by browsing on the polyps, by scooping mouthfuls from the hard stony material of living coral heads, or by feeding on

**"With increasing use of atolls for harbors, military bases, weapons-testing grounds, and tourist centers, more reef ecosystems are being damaged."**

branching coral tips. Today we know of 12 types of bony fish that feed on coral, representing a substantial part of the fish population in the neighborhood of a reef. An investigator estimated that 20 percent of the fish in East African seas could properly be called coral feeders.

Families of fish that include predators on coral are the spadefish, butterflyfish, damselfish, wrasses, surgeonfish, triggerfish, filefish, puffers, and porcupinefish. Families that might include predators are the sea chubs and the trunkfish. Several of the occasionally predatory families are those that are primarily herbivorous.

The major cause of natural destruction of reefs is the tropical storm; whole colonies can be uprooted and lashed ashore or carried out into deep water or broken up on the spot by furious wave action. And typhoons or hurricanes are accompanied by exceptionally heavy rains that lower the salinity of surface waters to a dangerous level.

A large area of lagoon coral growth was killed by "dark water" at the Dry Tortugas in 1878. Intense rainfall during a cyclone,

**"Tropical storms, not predators, are the main natural cause of the destruction of coral reefs."**

accompanied by a low tide, twice killed vast areas of corals on the coast of Australia. There have also been times when corals have died and the cause has remained unknown. The great glaciation cycles periodically change sea levels. In the course of geological times, it has been occasionally higher but often lower than today's mark (and at least once as low as minus 500 feet). Colonies are obviously killed when the sea level falls, but they have time to grow on the flanks of the rising islands.

With increasing use of atolls and other reef areas for harbors, military bases, airfields, weapons-testing grounds, and tourist attraction centers, more reef ecosystems are being seriously damaged. The classic case is that of Palmyra Atoll, where causeways built around the atoll perimeter during World War II prevented renewal of lagoon waters. The flourishing reefs died and were replaced by communities of algae.

*The flamingo tongue,* Cyphoma gibbosa, *pictured here on a purple sea fan, is among the many predators on the reef. The snail's soft spotted mantle completely covers its shell.*

# Crown-of-Thorns

The starfish known as crown-of-thorns (*Acanthaster planci*) is one of the most spectacular predators on coral reefs, and recently it has undergone population increases in many parts of the world. In some areas the crown-of-thorns is believed to be responsible for removing up to 90 percent of the living tissue from the hard corals of a reef.

At night the sea star settles on the stony coral, everts its entire stomach, and digests a complete polyp, leaving only the skeleton, while its 16 arms hold fast to the coral. Algae then settle into the empty skeleton, and borers soon follow.

In 1962-63 the multirayed starfish, a moving mound of spines about 15 inches across, became an infamous menace to stony corals of the Great Barrier Reef of Australia. By

*A living mound of spines,* the starfish called crown-of-thorns has caused considerable destruction on coral reefs all over the world.

1965 the crown-of-thorns, in a locustlike plague, had destroyed most of the large Green Island reef.

The removal by collectors of the triton shell, a predator on *Acanthaster planci,* might have been the cause of the recent outbreaks. Water pollution is another possible cause, and so is the testing of nuclear weapons. Blasting and dredging and even spearfishing have also been offered as causes.

*Crown-of-thorns sucks in the living polyp through its large digestive membrane, and leaves only the coral skeleton to be invaded by algae and borers.*

It has been suggested that the crown-of-thorns had not been a rarity before the recent plagues. If this is true, more harm than good could result from indiscriminate use of control measures. But if the crown-of-thorns plagues are not a new phenomenon, there is a possibility that human disturbances are increasing the frequency of the star by generating epidemics in areas where they might not have occurred naturally.

## Fighting Back

At the height of the devastation of the Great Barrier Reef, Australian authorities sent divers into the reef waters to locate and bury thousands of crown-of-thorns starfish. The destruction continued. A British scientific group based at Port Sudan in the Red Sea, after experiencing a reef invasion of the crown-of-thorns in 1970, increased their efforts to arrive at a solution. Undersea observations showed that the painted shrimp, triggerfish, and pufferfish were natural predators of *Acanthaster planci;* the local plague ended two years later, and it was concluded that it might have been natural. Several remedies were suggested, nevertheless. These included collection of the stars by divers, instructions to fishermen not to remove their natural enemies, and unfortunately the ridiculous action of injecting ammonium hydroxide into them.

Similar studies were made in Micronesia. American marine biologists, together with government agencies, examined all earlier conclusions—pollutants, blasting and dredging, shell collecting, spearfishing, and their effects on the predators of sea stars—and gave another possible explanation that typhoons were responsible for the redistribution of the food supply of predators on sea stars. This "environmental disturbance" idea now awaits experimental tests. Very recently another hypothesis was offered: that a chemical attraction among feeding stars might serve to bring them together.

The author's opinion is that *Acanthaster* is instrumental in reef destruction, but that its role has been vastly overemphasized.

*The triton (above), whose shell is a favorite of collectors, is a major predator on crown-of-thorns.*

*Injecting crown-of-thorns with poison (opposite) is a regrettable "remedy" that has been tried.*

## Other Reef Predators

Several species of fish destroy living corals either directly by feeding on the corallum or incidentally by breaking into coral colonies to expose animals dwelling within. These feeding activities are important factors in the bioerosive processes, as they retard coral growth and provoke the invasion of corals by alien growth through damaged surfaces. Parrotfish are the major consumers, in terms of biomass, of the algal material of the reef. The triggerfish, another major predator, bites the protruding surfaces to wrench loose large sections of the coral. The fish then searches through the fragments for bivalves, gastropods, crustaceans, and other things to eat. Some filefish, close relatives of the triggerfish, are very generalized predators on the reef, ingesting just about any form of benthic life they encounter. This includes such seemingly unwholesome things as sponges, gorgonians, hydroids, and stinging coral.

The preferred feeding sites of the parrotfish are exposed edges and the rims of platelike outgrowths. The feeding marks produced are usually nicks from one to three centimeters in length. These nicks are invaded by a variety of penetrating and creeping algae.

The most spectacular of the bulldozers of the reef are extremely large parrotfish called bumpfish by divers. They are ten to thirty pounds in weight, two to three feet in length, and have an enormous whitish protuberance on their forehead. They graze on the stone pastures in herds of one to ten dozen individuals, all day long. Each one of their bites takes away a piece of coral as large as an egg. They disperse at night to sleep in individual caves and regroup at dawn.

Erosion induced by coral predators can amount to as much as one-third of the annual coral growth.

*Cowries,* in these photographs, are gastropods with shiny, smooth shells. Some are known to feed upon coral, others are suspected of doing so.

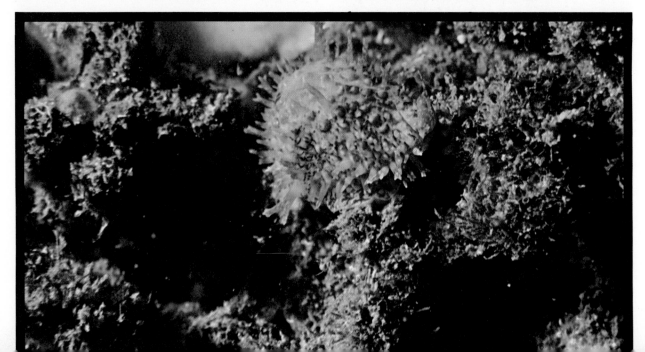

# Borers on Coral Reefs

Burrowing animals and plants cause a substantial part of the erosion of coral reefs. The borers include algae, molluscs, sipunculids, polychaetes, and certain sponges.

Green filamentous algae thread their way into the coral skeleton and can be found as a green zone within the surface of living colonies. One family of sponges are borers. They tunnel their way into coral heads, apparently to obtain a firmer anchor against the surge. They are known to make deep incisions in shallow water coral heads. Feather-duster worms also dig into coral to shelter their bodies and, when undisturbed, unfold their spirals in the water. Burrowing polychaetes feed like earthworms, passing large quantities of matter through their bodies. One polychaete grows to 16 inches in length. It is the palolo worm of the South Pacific. It passes the day deep within the reef and emerges at night to forage in the rocks for small invertebrates. Sipunculids, somewhat related to the earthworm, live below the surface of coral rubble and sediments. They swallow great quantities of the material of the reef to find the nutriment they need.

There are some borers among the reef's clam population. Some of them rasp away the limestone with the edges of their serrated shells. Some others dissolve the limestone with acid secretions. In Pacific reefs especially, bivalve borers do considerable damage.

If the population of borers was unchecked, it would destroy the reef even more efficiently than storms, but they are controlled by their own predators, such as triggerfish.

*Coral borers. Feather-duster worms (left and lower right) are commonly seen living on coral. Sponge (top right) has caused a deep crevice in the coral colony growing around it.*

# Chapter VIII. Man and the Coral Reef

Coral reefs are the most biologically productive of all natural communities, marine or terrestrial. Scattered over an area of more than 63 million square miles, they have traditionally supplied high-quality protein to native populations living near the sea in the tropics where other sources of protein are often inadequate. The reefs also act as a buffer against the ocean swell, protecting low tropical islands and atolls and thousands of miles of continental coastline. Man once

---

**"An awareness of the value and the fragility of coral and a concern for their protection is imperative."**

---

lived in relative harmony with that community. Today a thoughtless use of technology threatens to make that relation a destructive one.

One of the most serious dangers to the coral reefs is sedimentation. This threat is dramatically illustrated in Hawaii. Ancient Hawaiians cultivated taro in swampy lowlands and raised pigs and chickens. With the arrival of Westerners, sheep, goats, and cattle were introduced, and upland soil was plowed for sugar cane and pineapple. The impact of the resulting erosion has been tragic.

Since 1897 the shoreline of Molokai has advanced as much as a mile and a quarter across the reef flat. Elsewhere off Molokai, the reef is overlaid with four to 27 inches of red brown silt. Already 46 out of Hawaii's 70 species of birds are close to extinction.

In parts of the Great Barrier Reef of Australia, sedimentation is overtaking coral growth as a result of unplanned agriculture.

Most of the coral reefs have been destroyed by the erosion brought about by bulldozing

and surfacing of land that drains into Lindberg Bay off St. Thomas in the Virgin Islands. At Johnston Island in the Pacific great areas of the reefs have been destroyed as a result of the blasting and dredging of channels for the passage of ships.

Water containing a low concentration of salinity kills coral. This is evident in the Hawiian Islands where flooding is caused by the clearing of land for housing development.

Too much salinity caused by desalination plants is just as bad for corals. Thermal pollution by power plants threatens to become a serious problem, since tropical marine organisms live within only a few degrees of their upper thermal limits.

In areas that are polluted by sewage, certain algae invade corals, shutting them off from the necessary sunlight.

The use of dynamite to kill great masses of the fish and of bleach to flush them from their hiding places have become very widespread and very destructive practices. On a densely populated coast of Oahu in the Hawaiian Islands, after bleach was used by fishermen, almost no living animal remained.

Conservation regulations are urgently needed for coral reefs. The teaching of ecology in our schools is essential. An awareness of the value and the fragility of coral and a concern for their protection is imperative.

Damage done in but one day would take centuries to repair.

*In a coral park,* a diver pauses to admire the scenery. In the blue waters brilliant fish dart in and out of rocky crevices while exotic plant-animals wave languidly in response to the surge.

## Pilfering the Islands

By the end of the eighteenth century most of the coral islands of the Pacific had been "discovered" by Europeans. Explorers were followed by exploiters and developers.

While the ocean's whale populations were depleted, man looked to the shallow lagoons. There he found valuable mother-of-pearl shells to use for buttons and ornaments. Until the beds were exhausted, shell diving and collecting was one of the few income enterprises in the atolls of the eastern Pacific. The temporary hiatus caused by the outbreak of World War I saved the shell beds of the western Pacific from similar ruin.

Collecting of shells and corals (as well as spearing or capturing reef fish) remains one of the most formidable dangers to the reefs. Corals harvested for commercial use and for use in aquariums accounts for the destruction of many miles of reefs annually.

Early prospectors to the islands found another readily available crop at hand—natural fertilizer. Millennia had been required to cover the islands with guano, the powder-dry excrement of seabirds. In only a few years men scraped the famed "guano is-

lands" clean, loaded ships, and transported the fertilizer to the farms of America, Europe, and Australia. When none remained to be shoveled from the surface, it was ruthlessly crushed out of the rocks themselves.

More than any product of the islands, copra changed the lives of the native populations. Copra is the dried meat of the coconut, and it is a source of oil used for soap and for edible oils. To satisfy the world's copra demands, vast plantations were established on many islands in the South Pacific.

It is the affluent, jet-propelled tourist who may bring about the fifth collapse of the coral reefs. One by one the islands are being transformed into vacation centers. Channels for cruise ships are blasted through the reefs. Land is bulldozed for airports and building sites. Landfill is dredged from the sea. Untreated wastes poison the waters. Man's ill-conceived constructions could turn the coral castles into an archaeological tomb.

***Threats to the reefs.*** *Destructive by-products of unchecked development are channels dredged through living coral (above, left) and untreated waste (above, right) that upsets the fragile reefs.*

***Copra plantations,*** *like the one on opposite page, were established to satisfy the demand for coconut oil. They rob the soil of valuable nutrients.*

136

## War and Weapons Testing

World War II wreaked a great deal of destruction on many coral atolls in the South Pacific area. The material of war that still lies rotting on the islands and in their waters continues to do its damage. At Truk alone, over 30 Japanese warships and many airplanes and submarines lie strewn about at the bottom of the lagoon. The stockpiles of explosives left behind at war's end were subsequently used by the islanders to dynamite the reef for fish. Today there are few large fish in the area as a result of these practices.

The testing of nuclear weapons at coral atolls in the Pacific has also had a devastating effect. Three miles off Bikini in 180 feet of

*The sunken fleet.* Wrecked warships, submarines, and airplanes, casualties of World War II, continue to rust away in the quiet waters of Truk. They have provided shelter for marine life.

water are the hulls of the U.S.S. *Saratoga* and *Arkansas*, the Japanese battleship *Nagato*, and other ships that did not survive the atomic explosions of 1946. Oil from their fuel tanks still rises from them, polluting the surrounding water.

Radioactivity released by the blasts has been held within the atoll by concentration and reconcentration in the tissues of living things. Radioactivity was found to be present in the tissues of all species of fish sampled. The greatest amounts of radioactivity are in the fish that live chiefly on algae. In proportion to weight, it is higher in snails, crabs, and sea cucumbers than in fish and other vertebrates. Counts in algae are higher than in either vertebrates or invertebrates.

**Grim tombs.** *Softened by encrusting growths of corals, the twisted war machines sunk in Truk lagoon have been declared a monument to the men who died there, and are strictly guarded.*

# Preserving the Pharaohs

The visitor to a coral reef will be astounded by its beauty. Deep winding gullies carpeted with sand slice plateaus of coral so soft, so untouchable you fear they may fade away before your eyes. Graceful gorgonians raise arched branches like uplifted arms. Pastel-hued sea fans spread their lace to the eddying currents. Forests of staghorn coral, amazingly like antlers, crown the crest of the reef. Star coral, cactus coral, and leaf coral suggest decorations in a potentate's palace. Before your eyes pass the resplendent fish: sergeant majors in yellow with black bands; queen triggers, gray with two prominent blue stripes on the face; parrotfish arrayed in green, blue, purple, and even polka dots; unicorn filefish—studies in olive brown with black and white markings; horned cowfish; and others whose vivid colors combine all the hues of the rainbow and sunset. Such primitive beauty and solitude make a man feel he is trespassing on forbidden ground. The silence is awesome.

In Florida there is a great reef that runs west from Key Largo for about 220 miles. The reef was once in danger. Curio vendors were tearing it apart, using dynamite and crowbars. Bargeloads of corals, sponges, and shells were piled along the roadsides for sale to motorists. Fish collectors raided the waters, and spearfishermen stabbed everything that swam or crawled.

After much debate, legislation was finally passed that would protect at least 75 square miles of the reef. A preserve was created, to be administered by the governments of the United States and the State of Florida. Now in this one small corner of the pharaohs' realm, the work of these greatest of all builders can be enjoyed without fear of its being despoiled. It is a beginning, at least, and it is a hopeful sign.

# Keep the Stones Alive

Rooted deep in nothingness
Shored by dunes of lament
Gullied by solemn boulders
Stands the Wall of Walls.

> As high as hopes of the world
> As long as millions of centuries
> As thick as a nightmare
> As proud as a denial

The wall wails
Sobbing tears of sand,
Calcium cascades
For lakes of rock.

> Colossal tombstone
> In memory of birds and bees
> From sunken islands in the sun
> The wall is dying away.

On top, the narrow funeral wreath,
In a rainbow of throes and revels
Swarms with the tiniest masons
Stone flowers—stone pastures—
Stone cones and tubes and needles
Stones alive at the fringe of death.

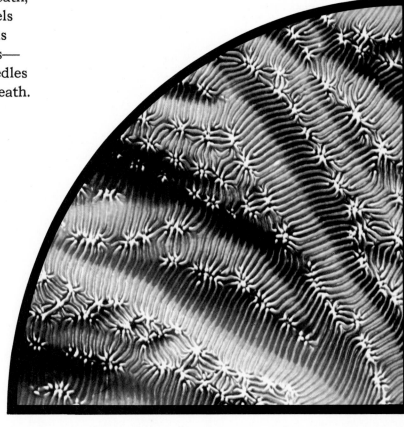

# Index

## ILLUSTRATIONS AND CHARTS:

Walter Hortens—50, 60; Howard Koslow—14-15, 30-31, 36-37, 90.

## PHOTO CREDITS:

Tony Chess—114 (bottom); J. P. Chevalier—22-23; Jim and Cathy Church—57, 72 (top); Patrick Colin—82; Ben Cropp—35, 42 (bottom), 127-128; David Doubilet—47, 69, 76, 77, 79, 88; Jack Drafahl—73, 118 (top); Cathy Engel—24; Bob Evans—72 (bottom); Freelance Photographers Guild: Ron Church—70 (bottom), Gordon DeLisle—40, James Dutcher—61, 71 (top), 80, 97, 99, 106, 131, 136, FPG—120 (top), Bob Gladden—2-3, 12-13 (bottom), 16-17, 48-49, 132, Bill Green—96, Jerry Jones—70 (top), Tom Myers—122-123 (bottom), Scott Penwarden—112-113, Steve Webster—133 (top), Western Marine Laboratories—120 (bottom); Henry Genthe—116 (bottom left); Al Grotell—12 (top), 54 (top); Bill McDonald—86-87 (bottom); T. J. Mueller—121; Dennis Muscatine—86 (top); NASA—21; D. R. Nelson—66-67 (right); Fred Roberts—83; Carl Roessler—25, 42 (top), 45, 55, 66 (left), 68, 92-93, 108, 114 (top left), 118 (bottom), 119; David Schwimmer—51; The Sea Library: Irwin Christen—20, Emerson Mulford—75 (bottom), 95, 130 (top), Valerie Taylor—38; Tom Stack & Associates: Ben Cropp—44, Keith Gillett—28-29 (bottom), 39, 43, E. R. Griffith—137, Dave La Touche—130 (bottom), Harold Simon—11; Taurus Photos: William J. Jahoda—125, C. Byran Jones—138 (bottom), 139, Dennis L. Taylor—53, Dave Woodward—46, 65, 90, 104-105, 107, 109, 110-111, 114-115 (top right), 133 (bottom),140-141; Ron Taylor—41, 81, 116 (top; bottom right); Valerie Taylor—103; Paul Tzimoulis—52, 56, 78 (bottom), 98, 126, 129, 135, 138 (top), 142; Myron Wang—89; World Life Research Institute: Devon Ludwig—62, 63.